Numerical and Quantitative Analysis

SURVEYS AND REFERENCE WORKS IN MATHEMATICS

Numerical and Quantitative Analysis

G Fichera
Università di Roma

Translated from Italian by
Sandro Graffi

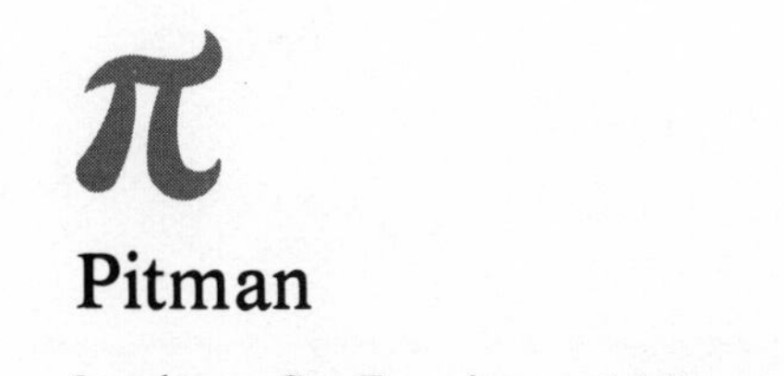

Pitman

London · San Francisco · Melbourne

PITMAN PUBLISHING LIMITED
39 Parker Street, London WC2B 5PB

FEARON–PITMAN PUBLISHERS INC.
6 Davis Drive, Belmont, California 94002, USA

Associated Companies
Copp Clark Ltd, Toronto
Pitman Publishing New Zealand Ltd, Wellington
Pitman Publishing Pty Ltd, Melbourne

First published 1974 by Accademia Nazionale dei Lincei, Rome
First Pitman edition 1978
Translated by S Graffi

AMS Subject Classifications: (main) 65–xx, 35–xx, 34–xx, 49–xx, 46–xx
(subsidiary) 45–xx, 31–xx, 73–xx

British Library Cataloguing in Publication Data
Fichera, Gaetano
Numerical and quantitative analysis. – (Surveys and reference works in mathematics; no. 3).
1. Differential equations – Numerical solutions
I. Title II. Series
515'.35 QA371

ISBN 0–273–00284–8

Reproduced and printed by photolithography
in Great Britain at The Pitman Press, Bath

To FRANCESCO G. TRICOMI
on his eightieth birthday

'. . . It is usually assumed by good mathematicians that numerical methods adapt some well known theory, sometimes not quite correctly, with the aim of obtaining more or less credible numerical results. Moreover, some of them are surprised to see that even under these circumstances it is often the case that no numerical results are produced. Instead, one finds in many papers a number of schemes preparing to attack numerically the problems, which never materialize, as if the conspirators have lost their heart . . . '

Alexander Weinstein

Preface

This survey aims to give a systematic presentation of the work undertaken by my students and myself during these last twenty years, in the field of Numerical, or, more generally, Quantitative Analysis. Although the main aim of this survey is to assemble as an organized development the results and methods of our research, it seemed worthwhile also to report on previous theories and results of other Authors to provide a suitable framework for the problems which are considered here and to facilitate the understanding of the results achieved. For instance, our results on Eigenvalue Theory could escape proper evaluation were they not compared with the existing literature and discussed in the light of the fundamental theories developed by Ritz, Weyl, Courant, Weinstein, Carleman, Agmon, and others.

For this reason I have given this survey much greater scope than merely adhering to a rigid account of my own work and that of my students.

A further aspect which I believe emerges from this survey is that of the 'philosophy' of approach to the subject of Numerical Analysis. The problems considered here and the methods which have been applied show, without the need of any further comment, the point of view adopted by the writer in the tradition of the School to which he belongs in dealing with problems of Numerical Analysis and in directing his students in this field.

Rome, November 1972

Gaetano Fichera

Preface to the English Translation

This survey originally appeared four years ago as a memoir of the Accademia Nazionale dei Lincei. In up-dating the text, three further Sections have been added, two in Part I and one in Part II. Thus, the survey refers also to some recent work carried out during these last four years.

I wish to express my most sincere thanks to Sandro Graffi for having taken the burden of translating the survey into English. My gratitude goes also to Alan Jeffrey for his careful revision of the English manuscript and for very useful comments and suggestions, and to Pitman for accepting to publish this book in its present fine printed form.

Rome November 1976 *Gaetano Fichera*

Contents

II *'A Priori'* Estimates for Solutions of Differential Equations

Part I: Computation of Eigenvalues

1. The Method Proposed by Picone for Computing Eigenvalues and Eigenvectors of General Linear Equations

Let S be a complex Hilbert space. Let us denote by (,) the *scalar product* and by $\| \ \|$ the *norm* in S.

Let V be a linear manifold in S and let E and M be linear transformations, defined in V and with range in S.

Consider in V the following eigenvalue problem:

$$\mathrm{E}u - \lambda \mathrm{M}u = 0, \qquad u \in \mathrm{V}. \tag{1.1.1}$$

Assume that the set of the eigenvalues of (1.1.1) has no (finite) accumulation point.

The following method for computing the eigenvalues and the eigenvectors of problem (1.1.1) was proposed by Picone. Let $\{u_k\}$ be a complete system in V, formed by linearly independent vectors. Let U_n be the linear manifold spanned by $u = \sum_{k=1}^{n} c_k u_k$ through the arbitrary variation of the complex constants $c_1, \ldots, c_n$.

Consider in U_n, for any fixed complex λ, the functional:

$$\mathrm{F}(u, \lambda) = \frac{\| \mathrm{E}u - \lambda \mathrm{M}u \|^2}{\| u \|^2}.$$

Let $\mu_n(\lambda)$ be the minimum of such a functional in $\mathrm{U}_n - \{0\}$ and $u^{(n)}(\lambda) = \sum_{k=1}^{n} c_k^{(n)}(\lambda)\, u_k$ be a minimizing vector. Let $\lambda_1^{(n)}, \lambda_2^{(n)}, \ldots, \lambda_{m_n}^{(n)}$ be all relative minima of $\mu_n(\lambda)$. Picone's method consists in assuming the numbers $\lambda_1^{(n)}, \lambda_2^{(n)}, \ldots, \lambda_{m_n}^{(n)}$ to be approximate eigenvalues of (1.1.1) and $u^{(n)}(\lambda_1^{(n)}), \ldots, u^{(n)}(\lambda_{m_n}^{(n)})$ to be the corresponding approximate eigenvectors.

Such a method has been developed and applied in papers by Nasta [1], Gröbner [2], Viola [3], Collatz [4], Kramers [5], Ghizzetti [6] and Fichera [7]. The most interesting numerical results are contained in [3], where the method has been applied to the following eigenvalue problem:

$$\frac{d^2}{dx^2}\left[(1-\vartheta x)^3 \frac{d^2 u}{dx^2}\right] - \lambda (1 - \vartheta x)\, u = 0, \tag{1.1.2}$$

$$u(0) = u(1) = u''(0) = u''(1) = 0,$$

ϑ being a constant such that $0 \leqslant \vartheta < 1$. When $\vartheta = 0$ the eigenvalues λ_k of (1.1.2) and the corresponding (normalized) eigenfunctions u_k are the following:

$$\lambda_k = k^4 \pi^4, \qquad u_k(x) = \sqrt{2} \sin k\pi x.$$

The method yields the following value as the first approximation to λ_1:

$$\lambda_1 \simeq \frac{3024}{31} \left(1 - \vartheta - \frac{\vartheta^2}{42}\right)$$

which, for $\vartheta = 0$, yields $\lambda_1 \simeq 97.548$, whereas the exact value is $\lambda_1 = \pi^4 = 94.4091 \ldots$

The following table shows the values given by the first three approximations for $\vartheta = 0$, compared with the exact values of the first three eigenvalues.

	First approximation	Second approximation	Third approximation	Exact values
λ_1	97.548	97.548	97.4091	97.4091
λ_2	–	1584.004	1584.004	1558.545
λ_3	–	–	7890.136	7890.13

(1.1.3)

The following table shows the approximations to the first eigensolution, given by the method, compared with its exact values.

	$\frac{1}{\sqrt{2}} u_1(x)$		$\sin \pi x$
	Second approximation	Third approximation	
0	0	0	0
0.1	0.308 190	0.309 003	0.309 017
0.2	0.583 080	0.587 644	0.587 785
0.3	0.798 279	0.808 599	0.809 017
0.4	0.934 938	0.950 353	0.951 057
0.5	0.981 748	0.999 160	1
0.6	0.934 938	0.950 353	0.951 057
0.7	0.798 279	0.808 599	0.809 017
0.8	0.583 080	0.587 644	0.587 785
0.9	0.308 190	0.309 003	0.309 017
1	0	0	0

(1.1.4)

The final table shows the results of the first four approximations to the eigenvalues $\lambda_1, \lambda_2, \lambda_3$ for the case $\vartheta = 0.5$.

	First approximation	Second approximation	Third approximation	Fourth approximation
λ_1	48.194	48.720	50.890	50.8129
λ_2	–	852.95	838.	836.9412
λ_3	–	–	–	4256.878

(1.1.5)

The mathematical justification of the method just described poses several problems.

(1) Establish under what conditions the set of eigenvalues of (1.1.1) has no accumulation point.

(2) In what sense the numbers $\lambda_1^{(n)}, \ldots, \lambda_{m_n}^{(n)}$ are to be considered approximate eigenvalues of (1.1.1).

(3) In what sense the vectors $u^{(n)}(\lambda_k^{(n)})$ $(k = 1, \ldots, m_n)$ are to be considered approximate eigenvectors of (1.1.1).

(4) After clarifying points (2) and (3), prove that the method 'converges'.

As far as point (2) is concerned, it is proposed in [3] to order the eigenvalues of (1.1.1) according to increasing modulus and, if the moduli coincide, according to increasing principal argument. This leads to an arrangement of the eigenvalues of (1.1.1) in a sequence: $\lambda_1, \lambda_2, \ldots, \lambda_k, \ldots$

After arrangement of the approximate eigenvalues $\lambda_1^{(n)}, \ldots, \lambda_{m_n}^{(n)}$, according to the same criterion, it is proposed in [3] to assume $\lambda_k^{(n)}$ as an approximation of λ_k and, consequently, $u^{(n)}(\lambda_k^{(n)})$ as an approximation of an eigensolution corresponding to λ_k. Unfortunately such a criterion is not acceptable. For example, if the two small white circles λ_1 and λ_2 of Fig. 1.1 indicate two eigenvalues of (1.1.1) and the small shaded circles the approximations given by the method, through the criterion proposed in [3], $\lambda_1^{(n)}, \lambda_1^{(n+1)}, \lambda_1^{(n+2)}, \ldots$ have to be assumed as approximations of λ_1 and $\lambda_2^{(n)}, \lambda_2^{(n+1)}, \lambda_2^{(n+2)}, \ldots$ as approximations of λ_2, whereas it is reasonable to think that $\lambda_1^{(n)}, \lambda_1^{(n+1)}, \lambda_1^{(n+2)}, \ldots$ approximate λ_2 and $\lambda_2^{(n)}, \lambda_2^{(n+1)}, \lambda_2^{(n+2)}, \ldots$ approximate the eigenvalue λ_1.

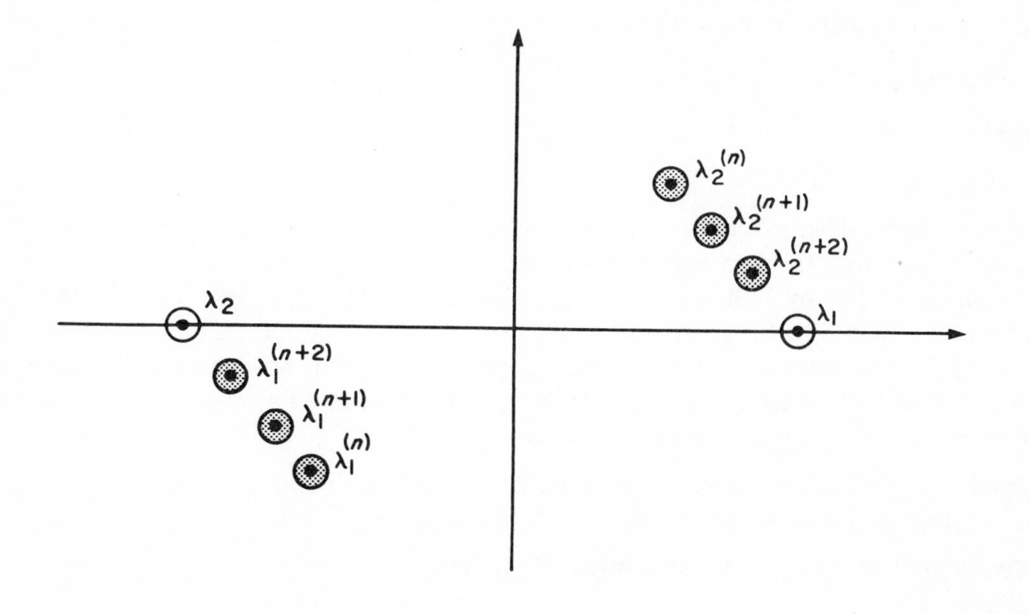

Fig. 1.1

As a matter of fact, the ordering chosen for the eigenvalues of the problem is an artificial one, since it is based upon the concept of principal argument, which corresponds to a convention among mathematicians and not to an intrinsic property of complex numbers.

Another major drawback of the method, as stated in [3], is due to the fact that the numbers $\lambda_1^{(n)}, \ldots, \lambda_{m_n}^{(n)}$, or a subset of them, may approximate no eigenvalue of the given problem.

Consider, indeed, the following case: let S be the space $\mathfrak{L}^2(0, \pi)$, V the manifold of the functions absolutely continuous with square summable first derivative in $(0, \pi)$, vanishing in $x = 0$. Let $\mathrm{E} \equiv \frac{\mathrm{d}}{\mathrm{d}x}$ and let M be the identity. Problem (1.1.1) becomes

$$\frac{\mathrm{d}u}{\mathrm{d}x} - \lambda u = 0, \qquad u(0) = 0 \tag{1.1.6}$$

The set of its eigenvalues has no accumulation point (it is empty!). Assume $u_k = \sin kx$. Then one has in U_n:

$$\mathrm{F}(u, \lambda) = \frac{\frac{2}{\pi} \sum_{k=1}^{n} (k^2 + |\lambda|^2) |c_k^2| - \sum_{h \neq k}^{1,n} \frac{hk}{h^2 - k^2} [1 - (-1)^{h+k}] (\lambda - \bar{\lambda}) c_h \bar{c}_k}{\frac{2}{\pi} \sum_{k=1}^{n} |c_k^2|} .$$

For any n the method yields some numbers $\lambda^{(n)}$, which are however meaningless, because they approximate no eigenvalue. In fact the set of the eigenvalues of problem (1.1.6) is empty. For example, if $\mathfrak{L}^2(0, \pi)$ is considered as a real space and hence λ and c_k are assumed real, one has:

$$\mu_n(\lambda) = (1 + \lambda^2)$$

and consequently

$$m_n = 1, \qquad \lambda_1^{(n)} = 0.$$

Hence the method would yield a (convergent!) sequence whose elements are all zero. On the other hand, the number zero is by no means an eigenvalue of problem (1.1.6).

The memoir [8] by Fichera, which inspired the subsequent notes [9] and [10] of Bassotti, is devoted to the mathematical justification of the method under discussion.

In [8] the theory is considered within the hypothesis that M is the identity operator I (we shall mention later the case in which M is assumed to be a more general operator) and that E satisfies the following assumption:

(a) E *is invertible and the inverse operator* $\mathrm{G} = \mathrm{E}^{-1}$ *is a compact operator* (*i.e. completely continuous*) *defined in* S *and with range in* V.

Problem (1.1.1) then takes the particular form

$$\mathrm{E}u - \lambda u = 0, \qquad u \in \mathrm{V}. \tag{1.1.7}$$

Such an hypothesis is verified by the eigenvalue problems that are encountered in applications. For example, let A be a bounded region of the cartesian space X^r and let

$$Eu = \sum_{h,k}^{1,r} a_{hk}(x) \frac{\partial^2 u}{\partial x_h \partial x_k} + \sum_{h=1}^{r} b_h(x) \frac{\partial u}{\partial x_h} + c(x)u \qquad (a_{hk} \equiv \bar{a}_{kh}) \tag{1.1.8}$$

with

$$\sum_{h,k}^{1,n} a_{hk}(x)\, \lambda_h \bar{\lambda}_k > 0 \qquad (|\lambda| > 0).$$

Let S be the space $\mathcal{L}^2(A)$ and V the space of functions having square summable first and second derivatives (space $H_2(A)$) satisfying the condition $u = 0$ on ∂A. If the coefficients a_{hk}, b_h, c are suitably smooth, we know, from the theory of 2nd order linear elliptic operators, that hypothesis (a) is satisfied. In fact we may assume G to be the Green operator relative to the elliptic operator (1.1.8) with Dirichlet boundary condition on ∂A. Coming back to the general case, the following statement can be made:

Hypothesis (a) *makes problem* (1.1.7) *such that the set of its eigenvalues has no accumulation point.*

It is stated in a precise way in [8] what sense is to be attributed to the convergence of the numbers given by the approximations towards the possible eigenvalues of the problem. Setting aside the hope of finding a non-artificial way of ordering the eigenvalues of a general operator E, such a convergence has to be intended in a *global sense*, i.e. as convergence of the *set* given by the approximations to the *set* of eigenvalues of problem (1.1.4). To this end it is convenient to introduce, in the space formed by all closed plane sets, a Hausdorff topology in the following way.

For a set C of the complex λ-plane we shall indicate by $I_\epsilon(C)$ its *envelope of radius* ϵ, i.e. the set of the points belonging to the plane, whose distance from C is less than ϵ. Now, if C is a closed set and ϵ and ρ two positive numbers, we shall define as a *neighbourhood* $\mathfrak{J}_{\epsilon\rho}$ C *of* C *with parameters* ϵ *and* ρ, the totality of the closed sets Γ satisfying the following conditions:

$$\Gamma \cap D_\rho \subset I_\epsilon(C \cap D_\rho), \qquad C \cap D_\rho \subset I_\epsilon(\Gamma \cap D_\rho),$$

where D_ρ stands for the circular disk $|\lambda| \leqslant \rho$.

One proves that the four Hausdorff axioms are satisfied:

(1) Every neighbourhood of C contains C.

(2) If Γ belongs to a neighbourhood of C, then there exists a neighbourhood of Γ completely contained in that neighbourhood.

(3) The intersection of two neighbourhoods of a same element C contains a neighbourhood of C.

(4) If C_1 and C_2 are distinct, there exists a neighbourhood of C_1 and a neighbourhood of C_2 having no common element.

The topological space just introduced will be indicated by $\mathcal{K}$. If Λ is the eigenvalue set of (1.1.7) and Λ_n the set given by the n-th approximation, *the convergence of* Λ_n *towards* Λ *is to be intended in the topology of the space* $\mathcal{K}$.

Of course, on the basis of what was remarked in the particular case of example (1.1.6), one cannot expect that Λ_n is formed by *all* numbers $\lambda_1^{(n)}, \ldots, \lambda_{m_n}^{(n)}$ but, more likely, by a subset of them, obtained by 'eliminating' from those numbers certain spurious ones which cannot be considered approximations of any eigenvalue. Such a 'sifting' is performed through the theorem we now expound.

Let us begin by supposing E to be such that $G^*(S) = V$, where G^* is the adjoint operator of G, G being the inverse operator of E. We remark that such an hypothesis is satisfied, for example, by the above-mentioned particular case, where E is given by (1.1.8). Assume, furthermore, $\{u_k\}$ to be such that $\{Eu_k\}$ is a complete system in S and $u_k \in G(V)$. The existence of such systems can be easily proved.

1.I. *Let* $\epsilon > 0$ *and* $\rho > 0$ *be arbitrarily fixed. Put*:

$$\eta_n(\lambda) = \frac{\|(E^* - \bar{\lambda} I)(E - \lambda I) u^{(n)}(\lambda)\|}{\|u^{(n)}(\lambda)\|}$$

(E^* = adjoint transformation of E).

Let $\Lambda_n(\rho)$ *be the set of the elements* $\lambda^{(n)}$, *chosen among* $\lambda_1^{(n)}, \ldots, \lambda_{m_n}^{(n)}$, *such that*

$$[\mu_n(\lambda^{(n)})]^2 \leqslant \sup_{\lambda \in D_\rho} [[\eta_n(\lambda)]^2 - [\mu_n(\lambda)]^2], \qquad |\lambda^{(n)}| < \rho. \tag{1.1.9}$$

The set $\Lambda_n(\rho)$ *is ultimately non-empty if the same is true for* $\Lambda \cap (D_\rho - \mathcal{F} D_\rho)$ *and ultimately one has*:

$$\Lambda_n(\rho) \subset \mathfrak{J}_{\epsilon\rho} \Lambda. \tag{1.1.10}$$

The meaning of this theorem is the following: suppose that we want approximations for all the eigenvalues of problem (1.1.7) such that $|\lambda| < \rho$, where ρ is arbitrarily fixed. Then, among the numbers $\lambda_1^{(n)}, \ldots, \lambda_{m_n}^{(n)}$ given by the n-th approximation and lying inside D_ρ, one has to choose those satisfying (1.1.9), and one has to neglect the other ones, which are to be considered as spurious numbers.

Through Theorem 1.I we deduce that, if $\lambda \in \Lambda$, then there exists a sequence $\{\lambda^{(n)}\}$, with $\lambda^{(n)}$ chosen among $\lambda_1^{(n)}, \ldots, \lambda_{m_n}^{(n)}$, such that:

$$\lim_{n \to \infty} \lambda^{(n)} = \lambda.$$

The following theorem holds:

1.II. *If the vector* $u^{(n)}(\lambda)$ *is unitary, the sequence* $\{u_n(\lambda^n)\}$ *is compact and all its compactness elements are eigenvectors of* (1.1.7) *corresponding to the eigenvalue* λ.

The methods of [8] have been continued in [9] and [10], where the case has been studied in which it is assumed that $U \equiv S$, $E \equiv I$, $M \equiv T$; T being a compact linear operator in the space S. In such a case the functional $F(u, \lambda)$ becomes:

$$F(u, \lambda) = \frac{\| \lambda T u - u \|^2}{\| u \|^2}.$$

If a system $\{u_k\}$ is chosen, which in the present case is supposed only to be complete in S, and if the numbers $\lambda_1^{(n)}, \ldots, \lambda_{m_n}^{(n)}$, are obtained, Theorems analogous to 1.I and 1.II hold.

In this case, for any fixed $\rho > 0$, the *selected set* $\Lambda_n(\rho)$ is obtained by choosing those $\lambda^{(n)}$ satisfying the following conditions:

$$|\lambda^{(n)}| < \rho, \qquad \mu_n(\lambda^{(n)}) \leqslant 1/2$$

$$[\mu_n(\lambda^{(n)})]^2 \leqslant \sup_{|\lambda| \leqslant \rho, \mu_n(\lambda) \leqslant 1/2} [\| \lambda T u^{(n)}(\lambda) + \bar{\lambda} T^* u^{(n)}(\lambda)$$

$$- |\lambda|^2 T^* T u^{(n)}(\lambda) \|^2 - 1 - \mu_n(\lambda)]^2].$$

Consider now the general problem (1.1.1) and, in addition to hypothesis (a), assume also that the following one is satisfied.

(b) *The operator* MG *is compact.*

Putting $\varphi = Eu$, problem (1.1.1) is equivalent to the following one:

$$\varphi - \lambda T \varphi = 0 \qquad \varphi \in S,$$

where T stands for MG.

This last problem has been studied in [9] and [10]. For example, hypothesis (b) is satisfied assuming E as indicated by (1.1.8), S and V in the way accordingly specified and

$$Mu = \sum_{h=1}^{r} \beta_h(x) \frac{\partial u}{\partial x_h} + \gamma(x) u$$

with $\beta_h(x)$ and $\gamma(x)$ continuous functions in $\bar{A}$. The theory can thus be applied to the following eigenvalue problem:

$$\sum_{hk}^{1,r} a_{hk}(x) \frac{\partial^2 u}{\partial x_h \partial x_k} + \sum_{h=1}^{r} b_h(x) \frac{\partial u}{\partial x_h} + c(x) u$$

$$- \lambda \left[\sum_{h=1}^{r} \beta_h(x) \frac{\partial u}{\partial x_h} + \gamma(x) u \right] = 0 \text{ in A}, \qquad u = 0 \text{ on } \partial A.$$

The method, although very general and nowadays fully justified as far as the selection of the values obtained (to be considered as actual approximate eigenvalues) is concerned, yields no indication about the error estimate, *that is, it yields no indication of the most important aspect in a numerical analysis method.*

The answer to the following question is not known: for any fixed $\epsilon > 0$ and any fixed $\rho > 0$, *how can an* $\eta(\epsilon, \rho)$ *be computed* such that (1.1.10) holds for $n > n(\epsilon, \rho)$?

As a matter of fact, this problem is, generally speaking, unsolved for any computational method for the (complex) eigenvalues of linear non-symmetric operators.

Some attempts in this direction have been made by Osborn [11], [12], [13] and, more recently, by Bramble and Osborn [14].

Osborn considers in S an operator like the operator E, satisfying hypothesis (a), but with G positive in the sense of operator theory, i.e. $(Gu, u) > 0$ for any $u \in S - \{0\}$. Since S is complex, this is known to imply that G is symmetric, i.e.

$$(Gu, v) = (u, Gv) \qquad (u, v \in S).^{(1)}$$

It follows that all eigenvalues of problem (1.1.7) are real and positive:

$$0 < \lambda_1 \leqslant \lambda_2 \leqslant \ldots \leqslant \lambda_n.$$

Osborn assumes that all the eigenvalues of problem (1.1.7) *are known together with the corresponding eigensolutions.* He then considers a continuous linear operator A defined in V, with range in S and *perturbs* through A the operator E, i.e. he considers the operator $\widetilde{E} = E + A$.

The following eigenvalue problem is proposed:

$$\widetilde{E}u - \lambda u = 0, \qquad u \in V. \tag{1.1.11}$$

At the basis of his treatment lies a theorem of Schwartz [15], found independently also by Gavurin [16], which states that:

1.III. *If the circular domains*

$$D_k : |\lambda - \lambda_k| \leqslant \| A \|$$

are pairwise disjoint, then in any D_k *there exists one and only one eigenvalue of problem* (1.1.11) *and there exist no eigenvalues of the problem outside these circles.*

Denoting by $\{u_k\}$ a complete orthonormal system of eigensolutions of the operator E, Osborn considers the operator

$$\widetilde{E}^{(n)} = \sum_{k=1}^{n} (\widetilde{E}u, u_k)u_k \tag{1.1.12}$$

whose eigenvalues are, of course, obtained by computing the eigenvalues of a matrix of rank n.

Assume that, in addition to the hypothesis of Theorem 1.III, the following one is also satisfied. Namely, $\| A \| < \lambda_1$. For a fixed $k < n$, let $\widetilde{\lambda}_k$ be the eigenvalue of problem (1.1.11) contained in D_k. Now, $\widetilde{\lambda}_1^{(n)}, \ldots, \widetilde{\lambda}_n^{(n)}$ being the (non-zero)

---

(1) Osborn also assumes that G has 'finite trace' in the Hilbert–Schmidt sense.

eigenvalues of the operator (1.1.12), Osborn shows that there exists an index i_k: $1 \leqslant i_k \leqslant n$ such that $|\tilde{\lambda}_k - \tilde{\lambda}_{i_k}^{(n)}|$ can be majorized by a quantity, vanishing with n^{-1} which Osborn explicitly computes. Unfortunately the way to determine such an index i_k, for a fixed k, is not shown.

Results of this type are continued in [13] and a new proof is provided for the case when E is an n-th order ordinary differential operator and A an ordinary differential operator of order $< n$.

Such results, whose theoretical interest is apparent, seem to be of limited value from the standpoint of actual numerical analysis.

Still less convincing are the results of [14]. As a matter of fact, they are based upon a general theorem, leaving considerable doubts about its effectiveness from the standpoint of numerical analysis. The authors of [14] consider the eigenvalue problem

$$\mathrm{T}u - \lambda u = 0, \qquad u \in \mathrm{S}; \tag{1.1.13}$$

T being a compact operator in S. T is assumed to satisfy certain conditions, usually satisfied by the Green operators of the differential operators. We need not quote these conditions here.

Let T_n be a sequence of compact operators, such that $\lim_{n\to\infty} \|\mathrm{T} - \mathrm{T}_n\| = 0$.

Let μ be a non-zero eigenvalue of T and let there be known a circular domain D centred on μ, in which there does not lie any other eigenvalue of T. By a theorem analogous to 1.III, if m is the (geometric) multiplicity of μ, then in D there will ultimately lie m and no more than m eigenvalues of T_n (counted according to their relative geometric multiplicities). Let these eigenvalues be $\mu_1^{(n)}, \ldots, \mu_m^{(n)}$. Now, the authors of [14] prove that there exist some index n_0 and some constant C, such that, for $n \geqslant n_0$

$$|\mu - \mu_k^{(n)}| \leqslant \mathrm{C}\alpha(\mathrm{T}, \mathrm{T}_n) \qquad (k = 1, \ldots, m)$$

where $\alpha(\mathrm{T}, \mathrm{T}_n)$ is some constant, explicitly computed, vanishing as $n \to \infty$. *What the values of n_0 and C are, is unknown.* In this way they claim to have determined 'the rate of convergence' of the eigenvalues of T_n towards those of T. Such a statement appears to be completely misleading. Take, for example, $\mathrm{S} \equiv \mathcal{L}^2(-\pi, \pi)$ and define T to be the operator

$$\mathrm{T}u = \int_{-\pi}^{\pi} \mathrm{K}(x, y)\, u(y)\, dy,$$

where

$$\mathrm{K}(x, y) = \sum_{k=1}^{\infty} e^{-k} \sin kx \cos ky + \frac{1}{\pi} \sin x \sin y + \frac{2}{\pi} \sin 2x \sin 2y.$$

For an arbitrarily fixed n_0, let

$$T_n u = \int_{-\pi}^{\pi} K_n(x, y)\, u(y)\, dy,$$

where

$$K_n(x, y) \begin{cases} = \sum_{k=1}^{n} e^{-k} \sin kx \cos ky + \frac{n}{\pi} \sin x \sin y + \frac{2n}{\pi} \sin 2x \sin 2y & \text{if } n < n_0, \\ = \sum_{k=1}^{n} e^{-k} \sin kx \cos ky + \frac{1}{\pi} \sin x \sin y + \frac{2}{\pi} \sin 2x \sin 2y & \text{if } n \geqslant n_0. \end{cases}$$

Then all hypotheses required in [14] are satisfied. The operator T has $\mu_1 = 1$ and $\mu_2 = 2$ as its (only) non-zero eigenvalues. The operator T_n has, as non-zero eigenvalues, $\mu_1^{(n)} = n$ and $\mu_2^{(n)} = 2n$ for $n < n_0$ and $\mu_1^{(n)} = 1$, $\mu_2^{(n)} = 2$ for $n > n_0$. Since n_0 may be taken as large as one wants, the convergence of the eigenvalues of T_n towards those of T may be delayed as much as one wants.

Bramble and Osborn can only claim to have 'ultimately' estimated the rate of convergence of $\mu_k^{(n)}$ towards μ. But the concept of an *ultimate* estimate (i.e. valid starting from some *unspecified* n_0) is, indeed, meaningless in numerical analysis!

2. Computation of the Eigenvalues of Symmetric Operators: Statement of the Problem

Let L be a linear operator with domain D_L contained in the Hilbert space S and with range in S. Let V be a linear manifold contained in D_L. The eigenvalue problem we shall consider is the following one:

$$Lv - \lambda v = 0, \qquad v \in V.^{(2)} \tag{1.2.1}$$

Problem (1.2.1) will be considered within the following hypothesis.

There exists a linear operator G, *mapping* S *in a subset of* V, *which*

(1) *is compact and strictly positive;*[3]

(2) *is such that* GL = LG = I (*identity operator*).

Through classical results of Hilbert space theory, it follows that the *characteristic set* (i.e. set of all eigenvalues) of the eigenvalue problem

$$Gu - \mu u = 0, \qquad u \in S \tag{1.2.2}$$

is a sequence of positive real numbers, each one with finite multiplicity, i.e. the set of

---

(2) This problem is analogous to (1.1.7). Here we are using different notations, in order to refer to the papers whose content is now under discussion.

(3) By G strictly positive we mean $(Gu, u) > 0$ for any $u \neq 0$. S being a complex Hilbert space, such a condition implies the symmetry of G. The compactness of G means that, if $\{u_n\}$ is a weakly convergent sequence in S, then $\{Gu_n\}$ is strongly convergent.

all eigensolutions corresponding to any single eigenvalue has finite dimension. Such a sequence converges towards zero. Let us indicate by

$$\mu_1 \geqslant \mu_2 \geqslant \ldots \geqslant \mu_k \geqslant \ldots$$

the sequence of the eigenvalues of (1.2.2), arranged in non-increasing order, each eigenvalue being repeated according to its multiplicity. Putting $\lambda_k = \mu_k^{-1}$, the sequence

$$\lambda_1 \leqslant \lambda_2 \leqslant \ldots \leqslant \lambda_k \leqslant \ldots$$

is the sequence of all eigenvalues of problem (1.2.1).

Our hypothesis embodies a simple formulation of the circumstance, usually occurring in the eigenvalue problems of mathematical physics, which consists in the possibility of translating into an integral equation, with a positive and symmetric kernel, the differential problem under investigation.

In the present exposition we shall confine our attention to the consideration of the following two questions.

(1) *Problem of the rigorous computation of the* λ_k*'s.*

By this we mean the problem of constructing, for any fixed k, two sequences $\{\tau_k^{(\nu)}\}$ and $\{\lambda_k^{(\nu)}\}$, such that

$$\tau_k^{(\nu)} \leqslant \lambda_k \leqslant \lambda_k^{(\nu)},$$

$$\tau_k^{(\nu)} \leqslant \tau_k^{(\nu+1)}, \qquad \lambda_k^{(\nu)} \geqslant \lambda_k^{(\nu+1)}, \qquad \lim_{\nu\to\infty} \tau_k^{(\nu)} = \lim_{\nu\to\infty} \lambda_k^{(\nu)} = \lambda_k.$$

The knowledge of the elements of both sequences $\{\tau_k^{(\nu)}\}$and $\{\lambda_k^{(\nu)}\}$ allows the computation of λ_k with an approximation error as small as one wants. The numbers $\tau_k^{(\nu)}$ will be called *lower bounds* (to λ_k) and the numbers $\lambda_k^{(\nu)}$ *upper bounds* (to λ_k).

(2) *Problem of the asymptotic distribution of the* λ_k*'s.*

Denote by $N(\lambda)$ the number of eigenvalues which do not exceed a bound λ. One has to construct in an explicit way a function $f(\lambda)$ which, for $\lambda \to +\infty$, has to be asymptotic to $N(\lambda)$, i.e. such that:

$$\lim_{\lambda\to\infty} \frac{N(\lambda)}{f(\lambda)} = 1,$$

in symbols: $N(\lambda) \simeq f(\lambda)$. If, as occurs in eigenvalue problems for elliptic differential operators, $f(\lambda)$ is of the type $c\lambda^q$, with $c > 0$, $q > 0$, putting $\lambda = \lambda_n$ and, hence, $N(\lambda_n) = n$, one obtains:

$$\lambda_n \simeq c^{-1/q}\, n^{1/q}$$

which describes the behaviour of λ_n as $n \to \infty$.

3. The Rayleigh–Ritz Method and the Plancherel Convergence Theorem

The problem of computing upper bounds $\lambda_k^{(\nu)}$ for λ_k was the first one to be solved, by the so-called Rayleigh–Ritz method, which, pioneered by Lord Rayleigh [17] in his celebrated treatise *The Theory of Sound* in the last decades of the last century, was later considered by Ritz in two memoirs [18], [19], which appeared in 1908and 1909.

The Rayleigh–Ritz method consists essentially in choosing a set of linearly independent vectors $\{v_i\}$, with $\{Lv_i\}$ complete in S, and, for any fixed integer number $\nu > 0$, in solving the secular equation:

$$\det\{(Lv_i, v_j) - \lambda\,(v_i, v_j)\} = 0 \qquad (i, j = 1, \ldots, \nu).$$

If the ν (all positive) roots of this equation are denoted by $\lambda_1^{(\nu)} \leqslant \lambda_2^{(\nu)} \leqslant \ldots \leqslant \lambda_\nu^{(\nu)}$, $\lambda_k^{(\nu)}$ is assumed as an upper bound to λ_k. It turns out that $\lambda_k^{(\nu)} \geqslant \lambda_k^{(\nu+1)}$ and, as Plancherel [20] first proved in 1923, one has:

$$\lim_{\nu\to\infty} \lambda_k^{(\nu)} = \lambda_k.$$

The Rayleigh–Ritz method is doubtless one of the most effective methods for numerical computations. Of course the rate of convergence of $\lambda_k^{(\nu)}$ towards λ_k depends on the choice of the system $\{v_i\}$, the choice to be made trying to take into account the peculiar characteristics of the particular problem under investigation.

It must be further remembered that the Rayleigh–Ritz method yields also an approximation procedure for the eigensolutions of the given problem (*see* [21], [22], [189]).

One has to remark that, while the existence of a complete system $\{v_i\}$ in V is trivial, since S and, hence, V are separable spaces, the actual construction of $\{v_i\}$ may offer some difficulty, depending on the definition of V.

4. The Maximum–Minimum Principle. Weinstein's Method for the Clamped Plate

Before treating the methods for constructing the lower bounds sequences $\tau_k^{(\nu)}$, it is convenient to recall a theorem, which bears fundamental importance in eigenvalue theory. We mean the theorem characterizing the so-called *maximum–minimum principle* for the eigenvalues. Assume $k > 1$. If $p_1, \ldots, p_{k-1}$ are arbitrary vectors in S, consider the minimum (whose existence can be proved) of the functional $\frac{(Lv, v)}{\|v\|^2}$ in the totality of the (non-zero) vectors v orthogonal to $p_1, \ldots, p_{k-1}$. Let such a minimum be $m(p_1, \ldots, p_{k-1})$. For $k = 1$, we shall consider the minimum of $(Lv, v)\,\|v\|^{-2}$ in $V - \{0\}$; it coincides with λ_1. If $k > 1$, the maximum of $m(p_1, \ldots, p_{k-1})$, when $p_1, \ldots, p_{k-1}$ vary arbitrarily in S, exists and coincides with λ_k. Hence, in symbols:

$$\lambda_k = \max \min_{\substack{v \perp p_i \\ i=1,\ldots,k-1}} \frac{(Lv, v)}{\|v\|^2}, \qquad v \in V. \tag{1.4.1}$$

Such a principle, observed by Fischer [23] for the case of symmetric matrices, was later proved for more general operators by Weyl [24], who identified all its interest in eigenvalue theory and based thereupon his asymptotic theory. Such a direction has been subsequently followed by Courant.

Notice that, if, for $k > 1$, as p_i an eigensolution relative to λ_i is assumed ($i = 1, \ldots, k-1$), one has:

$$m(p_1, \ldots, p_{k-1}) = \lambda_k. \tag{1.4.2}$$

It is easy to see that (1.4.2) holds not only with such a choice of the p_i's. The question, then, arises how to characterize any choice of $p_1, \ldots, p_{k-1}$, such that (1.4.2) holds. The problem has been solved by Weinstein [25], who however makes use of the explicit knowledge of any λ_n and of the corresponding eigensolutions.

In [26] (Th. 5.IV, page 345) a solution of the problem is pointed out, relative to particular classes of operators, based only upon the knowledge of the particular eigenvalue λ_k. It would however be extremely useful to obtain a solution of the problem not involving the knowledge of any λ_k and, moreover, of any eigensolution. Such a type of solution already appears, nevertheless, to be a question of remarkable difficulty in finite dimensional spaces.

The maximum–minimum principle is an essential tool in Weinstein's method [27] for constructing the sequences $\tau_k^{(\nu)}$. Such a method is the first effective contribution to the difficult problem of the lower approximation of the λ_k's. Although relative to a particular problem, it contains new ideas, which, as shown by Aronszajn [28], [29], can be extended to more general situations.

Weinstein considers the vibration problem of an elastic plate clamped along its boundary, i.e. the eigenvalue problem:

$$\Delta_2 \Delta_2 v - \lambda v = 0 \qquad \text{in A,}$$

$$v = \frac{\partial v}{\partial n} = 0 \qquad \text{on } \partial \text{A};$$

A is a bounded plane domain, $\Delta_2 \equiv \dfrac{\partial^2}{\partial x^2} + \dfrac{\partial^2}{\partial y^2}$ and, hence,

$\Delta_2 \Delta_2 = \dfrac{\partial^4}{\partial x^4} + 2\dfrac{\partial^4}{\partial x^2 \partial y^2} + \dfrac{\partial^4}{\partial y^4}$; $\dfrac{\partial}{\partial n}$ stands for the derivative along the normal direction to the boundary ∂A of A.

In the present case one has $L \equiv \Delta_2 \Delta_2$, where V is the linear manifold of the functions vanishing on ∂A together with their derivatives along the normal direction. The usual Hilbert scalar product for functions in A is denoted by (u, v). After some obvious integrations by parts, one has for (1.4.1):

$$\lambda_k = \max_{i=1,\ldots,k-1} \min_{v \perp p_i} \frac{\| \Delta_2 v \|^2}{\| v \|^2}, \qquad v \in \text{V}.$$

If we indicate by V^0 the manifold of the functions v such that $v = 0$ on ∂A, it follows from the maximum–minimum principle that, putting:

$$\tau_k^{(0)} = \max_{\substack{v \perp p_i \\ i=1,\ldots,k-1}} \min \frac{\| \Delta_2 v \|^2}{\| v \|^2}, \qquad v \in V^0,$$

one has for any k:

$$\tau_k^{(0)} \leqslant \lambda_k.$$

The numbers $\sigma_k = \sqrt{(\tau_k^{(0)})}$ are the eigenvalues of another classical problem, i.e. the one concerning the eigenfrequencies of a membrane spanning the region A and fixed along its boundary:

$$\Delta_2 v + \sigma v = 0 \quad \text{in A}, \qquad v = 0 \quad \text{on } \partial \text{A}. \tag{1.4.3}$$

On the other hand, by a remark which goes back to Zaremba [30], the manifold V can be characterized through the following conditions:

$$v = 0 \quad \text{on } \partial\text{A}; \quad \int_{\text{A}} \omega_s \Delta_2 v \, dx \, dy = 0 \quad (s = 1, 2, \ldots),$$

$\{\omega_s\}$ being a sequence of harmonic functions, complete in A; for example, if A is simply connected, the sequence comprises all harmonic homogeneous polynomials $\Re z^m$, $\Im z^m$ $(m = 0, 1, \ldots; z = x + iy)$.

Weinstein's idea consists in considering the *intermediate eigenvalues*:

$$\tau_k^{(\nu)} = \max_{\substack{v \perp p_i \\ i=1,\ldots,k-1}} \min \frac{\| \Delta_2 v \|^2}{\| v \|^2}, \qquad v \in V^\nu, \tag{1.4.4}$$

V^ν being defined by the conditions:

$$v = 0 \quad \text{on } \partial\text{A}; \quad \int_{\text{A}} \omega_s \Delta_2 v \, dx \, dy = 0 \quad (s = 1, 2, \ldots, \nu).$$

It follows immediately from the maximum–minimum principle that:

$$\tau_k^{(0)} \leqslant \tau_k^{(\nu)} \leqslant \tau_k^{(\nu+1)} \leqslant \lambda_k.$$

Aronszajn and Weinstein have in addition proved [31] that $\lim_{\nu \to \infty} \tau_k^{(\nu)} = \lambda_k$. The sequence $\{\tau_k^{(\nu)}\}$ is therefore a lower bounds sequence.

Of course what has so far been developed would be useless, unless one succeeds in implementing a procedure for the actual computation of the $\tau_k^{(\nu)}$'s. To this purpose Weinstein showed that, *if all eigenvalues and all eigenfunctions of problem* (1.4.3) (*called the base problem*) *are known, then the numbers* $\tau_k^{(\nu)}$ *of the sequence* $\{\tau_k^{(\nu)}\}$ *may be computed for any k.* Such numbers can be characterized not only through (1.4.4), but also as the eigenvalues of certain problems that are easy to describe, and which are called *intermediate problems*.

The question thus leads to the knowledge of the eigenvalues and eigenfunctions of problem (1.4.3) relative to the membrane.

If A is a rectangle, the eigenvalues and eigenfunctions of (1.4.3) being in this case known, Weinstein succeeds in obtaining through his method interesting numerical results, especially remarkable since they were obtained more than forty years ago, when the present powerful tools for numerical computation did not exist. We shall briefly mention the computation procedure for $\tau_k^{(\nu)}$, invented by Weinstein, but we prefer to do this, also for reasons of conciseness, in the framework of a more general situation, according to Aronszajn's setting, instead of in the particular case of the clamped plate.

5. Aronszajn's Formulation and Extension of the Weinstein Method

A general formulation of the Weinstein method is presented in [28] by Aronszajn. He proceeds in terms of problem (1.2.2) instead of (1.2.1), so that his aim is to construct an approximating sequence of upper bounds for any μ_k.

Suppose there exists a strictly positive compact operator Γ and an orthogonal projection P of S onto a subspace Ω of S,(4) such that

$$G = (I - P)\,\Gamma\,(I - P). \tag{1.5.1}$$

Notice that in the present case G, although positive, is not strictly positive; however its restriction to the orthogonal complement $\Omega^\perp$ of Ω(5) is strictly positive.

Suppose we are concerned with the computations of upper bounds to the eigenvalues of the following problem:

$$Gu - \mu u = 0, \qquad u \in \Omega^\perp.$$

---

(4) If Ω is a subspace (closed linear manifold) of S and u an element of S, there exists one and only one element Pu of Ω having minimal distance from u. Pu is defined to be the (*orthogonal*) projection of u onto Ω and P is the (*orthogonal*) *projector of* S *onto* Ω. The operator P is characterized by the following properties:

$$(Pu, v) = (u, Pv); \qquad P^2 = P.$$

If $\{\omega_k\}$ is an orthonormal complete system in Ω, one has:

$$Pu = \sum_k (u, \omega_k)\,\omega_k.$$

If $\{\omega_k\}$ is not orthonormal, but formed by linearly independent vectors, one has:

$$Pu = \lim_{n\to\infty} \sum_{hk}^{1,n} a_{hk}^{(n)} (u, \omega_h)\,\omega_k,$$

$\{a_{hk}^{(n)}\}$ being the inverse matrix of the $n \times n$ matrix $\{(\omega_i, \omega_j)\}$ $(i, j = 1, \ldots, n)$.

(5) The subspace of S formed by all vectors u such that $(u, \omega) = 0$, for any $\omega \in \Omega$ is called the *orthogonal complement* of Ω and is denoted by $\Omega^\perp$.

All the eigenvalues and all the eigenfunctions of the problem (called the *base problem*)

$$\Gamma u - \mu u = 0, \qquad u \in S \tag{1.5.2}$$

are assumed to be explicitly known, i.e. the spectral representation of Γ is known:

$$\Gamma u = \sum_{n=1}^{\infty} \mu_n^{(0)} (u, u_n) u_n. \tag{1.5.3}$$

Let $\{\omega_s\}$ be a system of linearly independent vectors, complete in the manifold Ω. Let Ω_ν be the ν-dimensional linear manifold spanned by $\omega_1, \ldots, \omega_\nu$ and let P_ν be the orthogonal projector of S onto Ω_ν. Put:

$$\Gamma_\nu = (I - P_\nu)\, \Gamma\, (I - P_\nu) \tag{1.5.4}$$

and let $\mu_k^{(\nu)}$ be the eigenvalues of the following problem (*intermediate problem of order* ν):

$$\Gamma_\nu u - \mu u = 0, \qquad u \in \Omega_\nu^{\perp}. \tag{1.5.5}$$

It follows from the maximum–minimum principle that:

$$\mu_k^{(0)} \geqslant \mu_k^{(\nu)} \geqslant \mu_k^{(\nu+1)} \geqslant \mu_k.$$

In addition, one proves that:

$$\lim_{\nu \to \infty} \mu_k^{(\nu)} = \mu_k. \tag{1.5.6}$$

The problem thus leads to the computation of $\mu_k^{(\nu)}$.

To this end a procedure may be applied analogous to the one already followed by Weinstein for the problem considered by him. Introducing the operator $R_\mu = (\Gamma - \mu I)^{-1}$, defined for any μ for which $\mu \neq 0$, $\mu \neq \mu_n^{(0)}$ ($n = 1, 2, \ldots$) by

$$R_\mu v = \frac{1}{\mu} \left[\sum_{n=1}^{\infty} \frac{\mu_n^{(0)}}{\mu_n^{(0)} - \mu} (v, u_n) - v \right],$$

it is necessary, following Weinstein, to distinguish the eigenvalues of (1.5.5) as being *non-persistent*, i.e. not belonging to the spectrum of Γ, or *persistent*, i.e. belonging to such a spectrum.

Remarking that one can put $\Gamma_\nu = \Gamma + D_\nu$, with

$$D_\nu u = \sum_{i,j}^{1,\nu} d_{ij}(u, v_i)\, v_j, \tag{1.5.7}$$

$\{d_{ij}\}$ being a non-degenerate Hermitian matrix of rank ν, by means of simple linear algebra arguments it can be proved that, denoting by $\{a_{ik}\}$ the inverse matrix of $\{d_{ij}\}$, the non-persistent eigenvalues of (1.5.5) are the zeros of the meromorphic function

$$W(\mu) \equiv \det \{(R_\mu v_j, v_h) - a_{jh}\} = 0 \qquad (j, h = 1, \ldots, \nu) \tag{1.5.8}$$

which do not belong to the spectrum of Γ. Every such a zero will have, as an eigenvalue of (1.5.5), a multiplicity equal to the complement of the rank of the matrix occurring in (1.5.8).

Now let μ be any eigenvalue of (1.5.2) and consider the operator

$$\widetilde{\mathrm{R}}_\mu v = \frac{1}{\mu}\left[\sum_{n=1}^{\infty}{}^{(\mu)} \frac{\mu_n^{(0)}}{\mu_n^{(0)}-\mu}(v, u_n) - v\right],$$

where $\sum_{n=1}^{\infty}{}^{(\mu)}$ means that those n for which $\mu_n^{(0)} = \mu$ must be omitted. Let $u^{(1)}, \ldots, u^{(s)}$ be a complete system of eigenvectors of (1.5.2) corresponding to the eigenvalue μ. Put:

$$\widetilde{\mathrm{W}}(\mu) = \det\left\{\begin{array}{c|c} (\widetilde{\mathrm{R}}_\mu v_j, v_h) - a_{jh} & (v_j, u^{(p)}) \\ \hline (u^{(p)}, v_h) & 0 \end{array}\right\} \tag{1.5.9}$$

$(j, h = 1, \ldots, \nu; p = 1, \ldots, s)$.

The way used for defining the $(\nu + s) \times (\nu + s)$ matrix on the right-hand side of (1.5.9) is obvious.

Now μ is a (persistent) eigenvalue of (1.5.5) if and only if it is a zero of the meromorphic function $\widetilde{\mathrm{W}}(\mu)$. Also in such a case its multiplicity, as an eigenvalue of (1.5.5), is the complement of the rank of the matrix defining $\widetilde{\mathrm{W}}(\mu)$. A computation procedure for $\mu_k^{(\nu)}$ has thus been implemented.

The above procedure may be regarded, with suitable modifications, as an abstract formulation of the original Weinstein method, relative to the clamped plates. The most important contribution by Aronszajn, however, is represented by an interesting modification of the procedure exposed so far. He notices indeed, basing his argument upon a remark which goes back to Weyl, that the inequality $\mu_k^{(0)} \geqslant \mu_k$ holds not only if Γ and G are related through (1.5.1), but also if one has:

$$\Gamma \geqslant \mathrm{G}. \tag{1.5.10}$$

Suppose now Γ and G are strictly positive and compact in S. As before, the eigenvalues of (1.5.2) are denoted by $\mu_k^{(0)}$, and those of the problem:

$$\mathrm{G}u - \mu u = 0, \qquad u \in \mathrm{S}$$

are denoted by μ_k.

Aronszajn [29] succeeds in pointing out a procedure in order to construct, for any ν, a compact operator Γ_ν such that $\Gamma_\nu = \Gamma + \mathrm{D}_\nu$, D_ν being hermitian and degenerate, i.e. representable under the form (1.5.7) and, furthermore, being such that:

$$\Gamma \geqslant \Gamma_\nu \geqslant \Gamma_{\nu+1} \geqslant \mathrm{G}.^{(6)}$$

(6) The construction of Γ_ν satisfying the required conditions may be performed, in addition to the way pointed out by Aronszajn, in several different ones. *See* in this context [32], ch. 17 and the references quoted therein.

Denoting then by $\mu_k^{(\nu)}$ the eigenvalues of the problem:

$$\Gamma_\nu u - \mu u = 0, \qquad u \in S,$$

one has (maximum–minimum principle):

$$\mu_k^{(0)} \geqslant \mu_k^{(\nu)} \geqslant \mu_k^{(\nu+1)} \geqslant \mu_k;$$

furthermore, D_ν may be constructed in such a way that (1.5.6) holds. On the other hand, in the procedure described above for computing $\mu_k^{(\nu)}$, use has not been made of the particular structure (1.5.4) of the operator Γ_ν, but only of the fact that $\Gamma_\nu = \Gamma + D_\nu$, D_ν being given by (1.5.7).

Hence, even in such a case, the meromorphic functions $W(\mu)$ and $\widetilde{W}(\mu)$ may be used for computing $\mu_k^{(\nu)}$. Notice that the validity of (1.5.1) does not imply the validity of (1.5.10); on the other hand, (1.5.10) does not imply the possibility of representing G according to (1.5.1). We shall come back later to the comparison between these two situations, to both of which the method, which seems to us is justified to call the Weinstein–Aronszajn method, can be applied.

What has been developed so far clearly shows that the greatest difficulty in applying the method consists in the need of disposing of a *base problem* (1.5.2) with an operator Γ, whose spectral representation (1.5.3) is explicitly known.

There exist several important examples for which a suitable base problem can be constructed, but it is also easy to exhibit other examples for which this construction cannot be done.

Another difficulty, encountered in applying the method, consists in the computation of zeros of meromorphic functions, such as $W(\mu)$ and $\widetilde{W}(\mu)$. It must be added, however, that Weinstein himself [27], [33], and later Bazley [34], have proposed some procedures in order to simplify, in some particular cases, the computation of the above-mentioned zeros, by means of special choices of the vectors occurring in the construction of D_ν. The problem was solved later in general by an ingenious idea of Weinberger [35], continued by Bazley and D. W. Fox [36]. It consists in modifying the above-mentioned methods, by replacing the *base operator* Γ with the *truncated operator*

$$\Gamma^{(m)} u = \sum_{n=1}^{m} \mu_n^{(0)} (u, u_n) u_n + \mu_{m+1}^{(0)} \left[u - \sum_{n=1}^{m} (u, u_n) u_n \right].$$

Hence, the name the *method of truncation* is attributed to the procedure.

The *intermediate operators* $\Gamma_\nu^{(m)}$ are always obtained from $\Gamma^{(m)}$ through a finite rank perturbation. In the present case, however, one has the advantage of obtaining the eigenvalues $\mu_k^{(\nu,m)}$ of $\Gamma_\nu^{(m)}$ as zeros of polynomials. Of course, in order to improve the (upper) approximations to the eigenvalues of G, it will be necessary to increase not only ν, but also m.

The Weinstein–Aronszajn method has been deeply studied and applied to several problems, by Bazley [34], [37], and by Bazley and D. W. Fox [36], [38], [39].

Applications of particular interest are those made to some eigenvalue problems occurring in theoretical physics. Let us remark that, in such a case, the relevant operators, although self-adjoint, are neither compact nor positive. Their spectrum consists, however, of a point component formed by an increasing sequence of eigenvalues, converging to a finite value and of a continuous component lying above such a value. The results of the method, as far as eigenvalue estimates are concerned, may be applied, through suitable modifications, also to this case; the theoretical questions relative to the convergence of the method, however, remain open.

The most conspicuous application consists perhaps in the result obtained by Bazley [37] in 1960, concerning the eigenvalues of the helium atom. If the Schrödinger operator for such an atom is denoted by H, one has:

$$\mathrm{H}u = -\frac{1}{2}\Delta_x u - \frac{1}{2}\Delta_y u - \frac{2u}{|x|} - \frac{2u}{|y|} + \frac{u}{|x-y|},$$

$\Delta_x (\Delta_y)$ being the Laplace operator relative to the variables x_1, x_2, x_3 (y_1, y_2, y_3) and $x(y)$ the point of coordinates x_1, x_2, x_3 (y_1, y_2, y_3). One has to compute the eigenvalues of the problem

$$\mathrm{H}u - \lambda u = 0, \tag{1.5.11}$$

u being a function of $|x|$, $|y|$ and $|x-y|$, such that

$$\int_{X^3}\int_{Y^3} |u|^2 \, dx_1 \, dx_2 \, dx_3 \, dy_1 \, dy_2 \, dy_3 < +\infty;$$

$X^3(Y^3)$ being the Cartesian space of the variables x_1, x_2, x_3 (y_1, y_2, y_3). The upper bounds for the first eigenvalues of (1.5.11) had been obtained by Kinoshita [40] through the Rayleigh–Ritz method (pushing the approximation up to order 80!).

Bazley [37] remarks that the operator

$$\mathrm{H}_0 u = -\frac{1}{2}\Delta_x u - \frac{1}{2}\Delta_y u - \frac{2u}{|x|} - \frac{2u}{|y|}$$

may be taken as a base operator for problem (1.5.11) and that all eigenvalues and eigenfunctions of H_0 can be explicitly computed. He proceeds then to apply the Aronszajn procedure for constructing the intermediate problems. By a suitable choice of the functions used to define the intermediate operators, he succeeds in computing the intermediate eigenvalues, thus arriving at excellent lower bounds to λ_1 and λ_2.

The results of Bazley (lower bounds) and Kinoshita (upper bounds) yield the following estimates (*see* [41]):

$$-3.063 \leqslant \lambda_1 \leqslant -2.9037; \qquad -2.165 \leqslant \lambda_2 \leqslant -2.145.$$

6. The Method of Weyl and Courant Concerning the Asymptotic Distribution of the Eigenvalues in the Classical Problems of Mathematical Physics

The great interest, from the point of view of physics, in solving problem (2) (asymptotic distribution of the eigenvalues), to which part of this exposition will be devoted, was stated very clearly by Lorentz at the beginning of this century and tackled first by Weyl, in his famous memoir [24], which appeared in 1912. Weyl first considers integral operators of Fredholm type and later proceeds to consider the eigenvalue problem concerning the Laplace operator Δ_2 or, more generally, a self-adjoint elliptic operator in two or three variables. In a subsequent paper he solves the problem of the eigenvibrations of a homogeneous and isotropic elastic body [42].

The method of Weyl was continued some years later by Courant, who refined and extended Weyl's results [43]. The investigations of Weyl and Courant are based upon the maximum–minimum principle, employed by Weyl in his memoir [24].

In order to give an idea of the type of arguments used by these authors, we shall consider the simple problem of the eigenfrequencies of a membrane fixed along its boundary, which leads to the eigenvalue problem:

$$\frac{\partial^2 u}{\partial x^2} + \frac{\partial^2 u}{\partial y^2} + \lambda u = 0 \quad \text{in A,} \tag{1.6.1}$$

$$u = 0 \quad \text{on } \partial \text{A},$$

A being a bounded domain of the xy plane. We shall now, essentially, follow the exposition of Courant (*see* [43], ch. VI).

Consider first the rectangular domain R: $0 < x < a$, $0 < x < b$ and therein the equation:

$$\frac{\partial^2 u}{\partial x^2} + \frac{\partial^2 u}{\partial y^2} + \lambda u = 0 \quad \text{in R,} \tag{1.6.2}$$

with the Dirichlet boundary condition

$$u = 0 \quad \text{on } \partial \text{R}, \tag{1.6.3}$$

or with the Neumann boundary condition:

$$\frac{\partial u}{\partial n} = 0 \quad \text{on } \partial \text{R}. \tag{1.6.4}$$

Equation (1.6.2) with the boundary condition (1.6.3) has the following eigenvalues:

$$\lambda_{hk} = \pi^2 \left(\frac{h^2}{a^2} + \frac{k^2}{b^2} \right), \tag{1.6.5}$$

with

$$h, k = 1, 2, \ldots,$$

whereas, if the boundary condition is given by (1.6.4), the eigenvalues are always given by (1.6.5), but one must then take:

$h, k = 0, 1, 2, \ldots.$

Denote by $N_R(\lambda)$ and $\widetilde{N}_R(\lambda)$ the number of eigenvalues of problems (1.6.2), (1.6.3) and (1.6.2), (1.6.4), respectively, which are not larger than λ. The determination of the asymptotic behaviour of $N_R(\lambda)$ and $\widetilde{N}_R(\lambda)$ corresponds to establishing the asymptotic behaviour, for $\lambda \to +\infty$, of the number of points with integer coordinates contained in the quarter ellipse:

$$\frac{x^2}{a^2} + \frac{y^2}{b^2} \leqslant \frac{\lambda}{\pi^2}, \qquad x \geqslant 0, y \geqslant 0,$$

excluding those points lying on the half-axes for the case (1.6.3) and including them for the case (1.6.4) (Fig. 1.2).

This is an old question, deeply investigated by experts in number theory, which is equivalent to the asymptotic determination of the number of vertices of a square mesh, contained in a fixed ellipse, when the mesh side goes to zero. A first simple answer to this question is provided by the following asymptotic relations:

$$N_R(\lambda) = \frac{ab}{4\pi}\lambda + \vartheta c\sqrt{\lambda}, \qquad \widetilde{N}_R(\lambda) = \frac{ab}{4\pi}\lambda + \widetilde{\vartheta}\widetilde{c}\sqrt{\lambda}, \tag{1.6.6}$$

c and $\widetilde{c}$ being constants independent of λ, $|\vartheta| < 1$ and $|\widetilde{\vartheta}| < 1$.

Let us come back now to problem (1.6.1) and let A be a bounded open plane domain, measurable in the sense of Peano–Jordan.

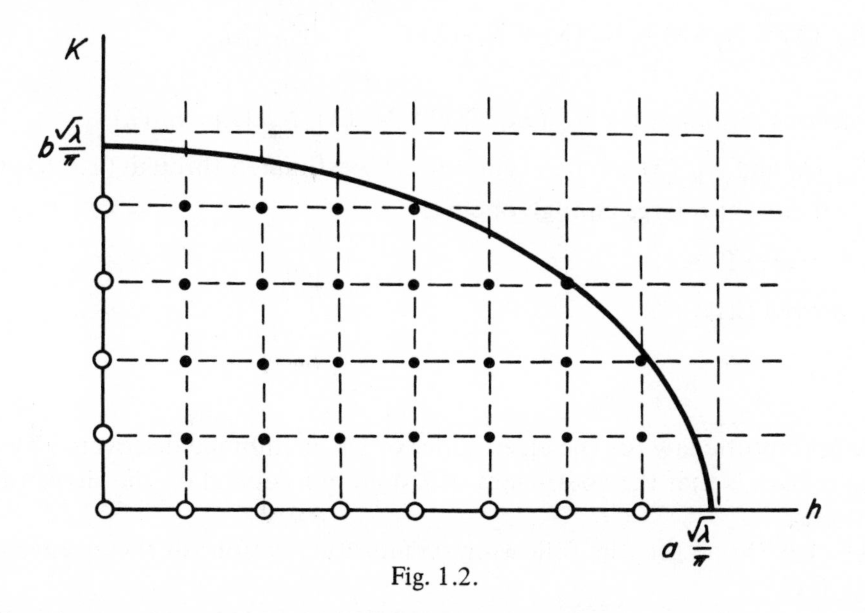

Fig. 1.2.

Introduce in the plane a lattice of side $1/n$, and let $R_1, \ldots, R_m, R_{m+1}, \ldots, R_{m+s}$ be all those squares having some common point with the closure $\overline{A}$ of A, and let among these squares be $R_1, \ldots, R_m$ all being internal to A (Fig. 1.3). Put:

$$P' = R_1 \cup \ldots \cup R_m, \qquad P'' = R_1 \cup \ldots \cup R_m \cup R_{m+1} \cup \ldots \cup R_{m+s}.$$

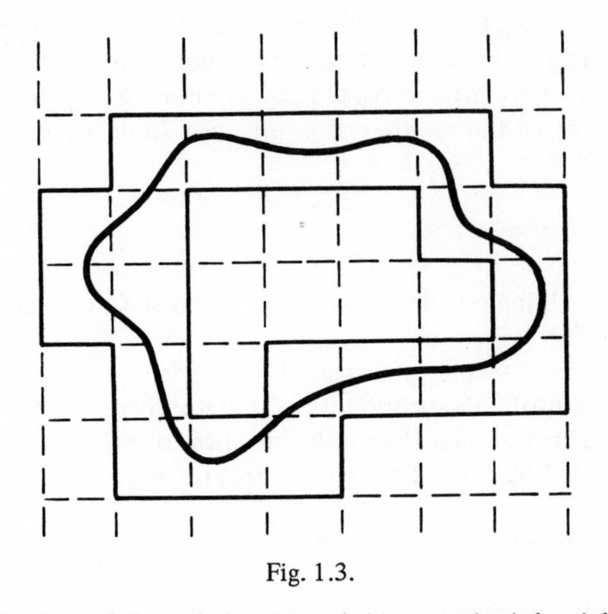

Fig. 1.3.

A repeated application of the maximum–minimum principle yields the following inequalities:

$$\sum_{i=1}^{m} N_{R_i}(\lambda) \leqslant N_{P'}(\lambda) \leqslant N_A(\lambda) \leqslant N_{P''}(\lambda) \leqslant \sum_{i=1}^{m+s} \widetilde{N}_{R_i}(\lambda),$$

with the obvious meanings for $N_{R_i}(\lambda)$, $N_{P'}(\lambda)$, $N_A(\lambda)$, $N_{P''}(\lambda)$, $\widetilde{N}_{R_i}(\lambda)$.

Since $N_{R_i}(\lambda)$ and $\widetilde{N}_{R_i}(\lambda)$ can be asymptotically estimated through (1.6.6) and since, for any $\epsilon > 0$ and for n large enough one has:

$$\text{area } P'' - \text{area } P' < \epsilon,$$

it is easily proved that

$$N_A(\lambda) \simeq \frac{\text{area A}}{4\pi} \lambda. \tag{1.6.7}$$

This is the asymptotic law for the eigenvalues of the membrane discovered by Weyl. An interesting remark is that the coefficient of λ does not depend on the shape of A, but only on its area.

Through (1.6.7) one gets the following asymptotic relation for the eigenvalue λ_n of problem (1.6.1):

$$\lambda_n \simeq \frac{4\pi}{\text{area A}} n. \tag{1.6.8}$$

Formula (1.6.7) (and hence (1.6.8)) is valid also if the Neumann boundary condition is considered, instead of the Dirichlet one, or, more generally, a boundary condition of the type:

$$\alpha u + \beta \frac{\partial u}{\partial n} = 0, \tag{1.6.9}$$

α and β being such as to ensure existence and uniqueness for the equation $\Delta_2 u = f$ with the boundary condition (1.6.9).

Formula (1.6.7), for such a general boundary problem, has been refined by Courant, who showed that

$$N_A(\lambda) - \frac{\text{area A}}{4\pi} \lambda = \mathcal{O}(\sqrt{\lambda} \log \lambda). \tag{1.6.10}$$

There exist a number of investigations directed to further improving (1.6.10), especially with the aim of giving a quantitative estimate to its right-hand side. Let us quote in particular the investigations of Pleijel [44] and Kac [45], recently extended and sharpened by McKean and Singer [46] in the context of r-dimensional problems for the Laplace–Beltrami operator on a Riemannian manifold.

Courant also considered the case of the biharmonic eigenvalue problem ([43], [47]):

$$\begin{aligned} \Delta_2 \Delta_2 u - \lambda u &= 0 \qquad \text{in A} \\ u = \frac{\partial u}{\partial n} &= 0 \qquad \text{on } \partial\text{A} \end{aligned} \tag{1.6.11}$$

This problem was later studied by Pleijel [48].

For such a problem they establish the asymptotic formula:

$$N(\lambda) \simeq \frac{\text{area A}}{4\pi} \sqrt{\lambda} \tag{1.6.12}$$

and hence:

$$\lambda_n \simeq \left(\frac{4\pi n}{\text{area A}} \right)^2. \tag{1.6.13}$$

7. The Carleman Method for the Asymptotic Theory of Eigenvalues and Eigenfunctions

A method completely different from the one of Weyl and Courant is developed by Carleman, in papers which appeared between 1934 and 1936 (*see* [49], [50]).

The Carleman method is particularly interesting because of its analytical content, which allows us to deduce information about the asymptotic behaviour not only of the eigenvalues, but also of the eigenfunctions. On the other hand, this method gave rise to almost all later investigations concerning the asymptotic theory of eigenvalues and eigenfunctions. Let us try to give an idea of the Carleman method, by considering a particularly simple problem, i.e. the analogue of (1.6.1), considered however in a bounded domain A of the cartesian space X^3 of the variables x_1, x_2, x_3:

$$\begin{cases} \Delta_2 u + \lambda u = 0 & \text{in A,} \\ u = 0 & \text{on } \partial\text{A.} \end{cases} \tag{1.7.1}$$

A will be assumed to possess a sufficiently smooth boundary.

The analytical tool, upon which the Carleman method is based, is not, as for the method of Weyl and Courant, the maximum–minimum principle, but a tauberian theorem due to Hardy and Littlewood [51] (*see* also [52]). It states that, if $\sigma(\lambda)$ is a non-decreasing function, defined for $\lambda \geqslant 0$ and if, for any $t > 0$, the Stieltjes integral

$$\int_0^{+\infty} \frac{\mathrm{d}\sigma(\lambda)}{\lambda + t} \tag{1.7.2}$$

exists and if, for $t \to +\infty$,

$$\int_0^{+\infty} \frac{\mathrm{d}\sigma(\lambda)}{\lambda + t} = ct^{\alpha-1} + o\,(t^{\alpha-1}), \tag{1.7.3}$$

with $c \geqslant 0$ and $0 < \alpha < 1$, then, for $\lambda \to +\infty$, one has:

$$\sigma(\lambda) = c\,\frac{\sin \pi\alpha}{\pi\alpha}\lambda^{\alpha} + o\,(\lambda^{\alpha}).$$

Let $x \equiv (x_1, x_2, x_3)$, $y \equiv (y_1, y_2, y_3)$. Let $G_\lambda(x, y)$ be the Green function of the problem $\Delta_2 u + \lambda u = f$ in A, $u = 0$ on ∂A, which exists for any λ different from any eigenvalue λ_k. $G_\lambda(x, y)$ is given by

$$G_\lambda(x, y) = \sum_{k=1}^{\infty} \frac{u_k(x)\,u_k(y)}{\lambda_k - \lambda}, \tag{1.7.4}$$

$u_k(x)$ being an eigenfunction of (1.7.1) corresponding to λ_k. The expansion (1.7.4) is justified in the sense of the $\mathfrak{L}^2$ convergence in A $\times$ A. If one assumes $t > 0$ and $\lambda = -t$, then one may write:

$$G_{-t}(x, y) = \frac{\exp(-\sqrt{t}\,|x-y|)}{4\pi\,|x-y|} - \gamma_t(x, y),$$

$\gamma_t(x, y)$ being the well known *regular part* of the Green function. Hence one obtains:

$$\begin{aligned} G_{-t}(x, x) - G_0(x, x) &= -\frac{\sqrt{t}}{4\,\pi} - \gamma_t(x, x) + \gamma_0(x, x) \\ &= -t \sum_{k=1}^{\infty} \frac{|u_k(x)|^2}{\lambda_k(\lambda_k + t)} \end{aligned} \tag{1.7.5}$$

this last series being uniformly convergent in A.

Using the maximum principle for the solutions of $\Delta_2 u - tu = 0$, it is easily seen that

$$\gamma_t(x, x) = \mathcal{O}\{\exp[-\sqrt{t}\,\delta(x)]/\delta(x)\},$$

$\delta(x)$ being the distance between x and ∂A.

From (1.7.5), after integration over A, one gets:

$$\sum_{k=1}^{\infty} \frac{1}{\lambda_k(\lambda_k + t)} = \frac{\text{vol A}}{4\pi} t^{-1/2} + o(t^{-1/2}). \tag{1.7.6}$$

On the other hand, introducing the function

$$\sigma(\lambda) = \sum_{\lambda_k \leqslant \lambda} \frac{1}{\lambda_k},$$

the left-hand side in (1.7.6) may be rewritten in the form (1.7.2), so that (1.7.6) reduces to the type of condition occurring as hypothesis in the tauberian theorem of Hardy and Littlewood. Therefore it follows, for $\lambda \to +\infty$:

$$\sigma(\lambda) = \frac{\text{vol A}}{2\pi^2} \lambda^{1/2} + o(\lambda^{1/2}).$$

Taking into account the fact that

$$\text{N}(\lambda) = \int_0^{\lambda} l \, d\sigma(l) = \sigma(\lambda) - \int_0^{\lambda} \sigma(l)\, dl,$$

we obtain

$$\text{N}(\lambda) = \frac{\text{vol A}}{6\pi^2} \lambda^{3/2} + o(\lambda^{3/2}).$$

Let us remark that if, instead of using (1.7.6), we have applied the foregoing procedure directly to (1.7.5), we would have obtained results relative to the eigenfunction's asymptotic behaviour.

Carleman applied his procedure to several situations, including also not necessarily self-adjoint operators. A number of authors have continued Carleman's investigations. Pleijel extended the Carleman method to equations of order higher than the second, recovering in the way mentioned above, a rigorous proof of the asymptotic law for the eigenvalues of the clamped plate [48], already considered by Courant. Keldys [53] considered non self-adjoint problems. Gårding [54], [55], immediately followed by Browder [56] (*see* also [57]), studied the case of the Dirichlet problem for a general $2m$-th order self-adjoint elliptic operator. Further results are due to Ehrling [58], Odhonoff [59], Bergendal [60]. It was, however, Agmon [61], [62], [63], [64], [65], [66], who formulated a complete theory, extending the Carleman one to the most general eigenvalue problems for higher order elliptic operators. He makes use, in addition to the theorem of Hardy and Littlewood, of even more sophisticated analytical tools, such as, for example, a recent tauberian theorem due to Malliavin [67].

Let us briefly mention some results obtained by Agmon. If $\xi \equiv (\xi_1, \ldots, \xi_r)$ is a real r-component vector and α a multi-index, i.e. an ordered set of r non-negative integers, put:

$$\xi^\alpha = \xi_1^{\alpha_1} \xi_2^{\alpha_2} \ldots \xi_r^{\alpha_r},$$

where the convention has been adopted of assuming $\xi_k^{\alpha_k} = 1$ if $\xi_k = 0$, $\alpha_k = 0$. The symbol D^α stands for the partial derivative:

$$D^\alpha = \frac{\partial^{\alpha_1 + \ldots + \alpha_r}}{\partial x_1^{\alpha_1} \ldots \partial x_r^{\alpha_r}}.$$

For any multi-index α we will put, as of common use now, $|\alpha| = \alpha_1 + \ldots + \alpha_r$. Consider in the space X^r the $2m$-th order differential operator, with real and suitably smooth coefficients:

$$L(x, D) = \sum_{|\alpha| \leq 2m} a_\alpha(x) D^\alpha.$$

It will be supposed *formally self-adjoint*, i.e., for any u of class C^{2m}, let

$$\sum_{|\alpha| \leq 2m} a_\alpha(x) D^\alpha u = \sum_{|\alpha| \leq 2m} (-1)^{|\alpha|} D^\alpha [a_\alpha(x) u].$$

Let the operator $L(x, D)$ be *positive elliptic*, i.e., putting

$$L_0(x, D) = \sum_{|\alpha| = 2m} a_\alpha(x) D^\alpha,$$

for any $\xi \neq 0$ and for any x, one has:

$$L_0(x, \xi) > 0. \tag{1.7.7}$$

Consider, in the bounded domain A of X^r, the equation:

$$L(x, D) u + (-1)^{m-1} \lambda u = 0$$

associated with suitable boundary conditions, such that the resulting boundary value problem has a Green operator compact and self-adjoint in the sense that the relative Green function is symmetric. Let $\{\lambda_k\}$ be the sequence of the eigenvalues of the problem, which can be positive and negative. For any $\lambda > 0$ denote by $N^+(\lambda)$ the number of all non-negative eigenvalues not exceeding λ and by $N^-(\lambda)$ the number of all negative eigenvalues not less than $-\lambda$. Set

$$\gamma = \frac{1}{(2\pi)^r} \int_A dx \int_{L_0(x,\xi) < 1} d\xi.$$

The following asymptotic relations have been proved by Agmon:

$$N^{+}(\lambda) = \gamma\lambda^{r/2m} + \mathcal{O}(\lambda^{r/2m}), \qquad N^{-}(\lambda) = \mathcal{O}(\lambda^{r/2m}). \tag{1.7.8}$$

Notice that (1.7.8) shows that the positive ellipticity condition implies that the positive eigenvalues are *more* than the negative ones. However it may very well happen, as shown by Agmon [62], through an example, that there are cases with infinite positive eigenvalues and infinite negative ones.

Before concluding this Section, we wish to mention a further procedure for the asymptotic study of eigenvalue problems, proposed by Minakshisundaram [68] between 1948 and 1949. The purpose of such a procedure was to introduce into eigenvalue problems methods and concepts used in number theory, in order to get the analogues of the theorems concerning the asymptotic distribution of prime numbers.

Considering problem (1.6.1), Minakshisundaram introduces, following Epstein, an analogue of the Riemann ζ function and an analogue of the ϑ function. After having established the functional relation, corresponding to the classical one between ζ and ϑ, Minakshisundaram, following Riemann, deduces thereby certain fundamental properties for the analogues of the function ζ considered by him. In such a way he recovers the asymptotic laws relative to the eigenvalues and eigenfunctions of the membrane.

8. Preliminaries to the Orthogonal Invariants Method

Let us consider again problem (1) posed in Section 2 (rigorous computation of the eigenvalues), in order to report on some results obtained in these last years. There appear to be contributions both from the theoretical standpoint as well as from the numerical one, to the problem of computing lower bounds to the eigenvalues λ_k of problem (1.2.1).(7)

It is helpful, we think, to recall the considerations which led to the formulation of a method for computing lower bounds to the λ_k's, called the *orthogonal invariants method.*

Let us consider problem (1.2.2). In order to implement a procedure for the computation of the μ_k's, the operator G may, of course, be replaced by another arbitrary one T, compact and strictly positive on a Hilbert space Σ, provided T has the same eigenvalues as G, with the same multiplicities. It is, then, spontaneous to define G and T *equivalent* with respect to problem (1). We deal here, as is quite obvious, with an actual equivalence concept, since the symmetry, transitivity and reflexivity axioms are immediately verified.

A natural question is to find an analytical expression of this equivalence relation. This question is easily answered by proving that T and G are equivalent in the above-mentioned sense, if and only if there exists a unitary operator U from the Hilbert space Σ onto the space S, such that $T = U^{-1}GU$. Hence the equivalence concept we have introduced is nothing other than the well known *unitary equivalence* between two operators of the Hilbert spaces S and Σ. We can thus say that the sequence $\{\mu_k\}$ is a

---

(7) *See* [26], [32], [69], [70], [71], [72], [73], [74], [75], [76], [77].

complete system of invariants with respect to the unitary equivalence or, more shortly, a *complete system of orthogonal invariants* for compact and strictly positive operators.

The following questions then arise:

(1) Since the orthogonal invariants of system $\{\mu_k\}$ are difficult to construct (our problem is exactly the computation of the μ_k's!), is it possible to replace $\{\mu_k\}$ by another sequence of orthogonal invariants, explicitly computable, in such a way as to characterize an equivalence class with respect to the unitary group?

(2) Assuming such a new system of orthogonal invariants to have been constructed, is it possible to achieve thereby the computation of the μ_k's?

In this Section we will answer question (1) and in the next one question (2).

The problem of constructing complete systems with respect to the unitary equivalence between two bounded operators on Hilbert spaces was investigated first by Hellinger [78]. His investigations have been later continued by Hahn [79], and by Plessner and Rohlin [80]. In more recent times Halmos [81] has considered questions of this type. The results obtained by these authors, however, are not suitable for our purposes, both because they concern much more general operators than those we are interested in and because they are directed towards aims that are different from ours.

We shall, on the contrary, try, although within some particular classes of positive and compact operators, to construct an orthogonal invariants theory, which, directly generalizing the classical one concerning finite dimensional spaces, makes problem (2), as posed before, solvable.

For any compact and strictly positive G and for any integer $n > 0$, we will set, denoting by $v_1, \ldots, v_s$ s arbitrary vectors in S:

$$G^{(n)}(v_1, \ldots, v_s) = \det(G^n v_i, v_j) \qquad (i, j = 1, \ldots, s).$$

Let $\{v_k\}$ $(h = 1, 2, \ldots)$ be an orthonormal complete set of vectors in S. Let $n > 0$ and $s > 0$ be two integer numbers. Set $\mathfrak{J}_0^n(G) = 1$ and, for $s > 0$,

$$\mathfrak{J}_s^n(G) = \frac{1}{s!} \sum_{k_1 \cdots k_s} G^{(n)}(v_{k_1}, \ldots, v_{k_s}) \tag{1.8.1}$$

where the sum occurring in the right-hand side series is extended over all possible choices of s positive integer indices. This series will have a finite or infinite sum, its terms being non-negative. The sum will not depend on the choice of the summation procedure. It is immediately verified that

$$\mathfrak{J}_s^n(G^m) = \mathfrak{J}_s^{nm}(G).$$

The following theorems hold (*see* [26] and [32]).

1.IV. *$\mathfrak{J}_s^n(G)$ does not depend on the particular system $\{v_k\}$, chosen in order to define it; this is equivalent to saying that $\mathfrak{J}_s^n(G)$ is an orthogonal invariant of* G.

1.V. *It turns out that $\mathfrak{J}_s^n(G) < +\infty$ if and only if $\mathfrak{J}_1^n(G) < +\infty$.*

Such a theorem leads to the introduction of the class $\mathfrak{C}^n$ of all compact and strictly positive operators G, such that $\mathfrak{J}_1^n(G) < +\infty$. One proves that $\mathfrak{C}^n$ is characterized by the fact that any $G \in \mathfrak{C}^n$ is such that the 'trace' of G^n is finite. For $m < n$ it turns out that $\mathfrak{C}^m \subset \mathfrak{C}^n$. The operators G, we shall consider from now on, will always be assumed to belong to $\mathfrak{C}^n$, for some n that is large enough. This is an actual restriction of the theory that is to be developed since there exist compact and strictly positive operators not belonging to any class $\mathfrak{C}^n$. The restriction, however, does not exclude all those operators, usually occurring in boundary problem theory for elliptic differential equations.

The following theorems establish the monotonicity and continuity properties of the orthogonal invariants introduced above.

1.VI. *If* $G_1 \leqslant G_2$, *then* $\mathfrak{J}_s^n(G_1) \leqslant \mathfrak{J}_s^n(G_2)$.

1.VII. *If, for* $k \to \infty$, G_k *converges uniformly to* G (*i.e.* $\lim_{k\to\infty} \|G_k - G\| = 0$) *and* $G_k \leqslant G_0$ (G_0 *positive compact operator*), *then one has*: $\lim_{k\to\infty} \mathfrak{J}_s^n(G_k) = \mathfrak{J}_s^n(G)$.

The principal theorem for the orthogonal invariants theory is represented by the following:

1.VIII. *The sequence* $\{\mathfrak{J}_s^n(G)\}$ ($s = 1, 2, \ldots$) *forms a complete system of orthogonal invariants with respect to the unitary equivalence of two operators, belonging to* $\mathfrak{C}^n$.

An answer has thus been given to question (1) posed earlier, through the construction of the required sequence of orthogonal invariants, each one of them being explicitly defined by (1.8.1).

We shall come back in a moment to the possibility of computing $\mathfrak{J}_s^n(G)$ by means of (1.8.1). First let us remark that the previous theory, essentially represented by Theorems 1.IV, 1.V, 1.VI, 1.VII and 1.VIII, allows us to answer a question posed in Section 5, about the comparison between two situations, expressed by (1.5.1) and (1.5.10) respectively, to both of which the Weinstein–Aronszajn method can be applied. These two situations are, indeed, to be compared, referring them not to the particular operators under investigation, but considering each operator inside the equivalence class (with respect to the unitary group), to which it belongs. Starting the problem in this way, it is realized that the operator $(I - P)\Gamma(I - P)$, occurring in the right-hand side of (1.5.1), is, considered in $\Omega^\perp$, unitary equivalent to the following operator (restricted to the orthogonal complement of its kernel):

$$\Gamma - \Gamma^{1/2} P \Gamma^{1/2};$$

$\Gamma^{1/2}$ stands for the positive square root of the operator Γ. On the other hand, since one has $\Gamma \geqslant \Gamma - \Gamma^{1/2} P \Gamma^{1/2}$, the situation (1.5.1) can be related to (1.5.10). This means that the case considered by Aronszajn is more general than the one stating, in an abstract way, the situation originally studied by Weinstein.

The above-mentioned unitary equivalence between $(I - P)\Gamma(I - P)$ and $\Gamma - \Gamma^{1/2} P \Gamma^{1/2}$ may be demonstrated by making a suitable use of Theorem 1.VIII.

From the theoretical standpoint it is interesting to establish the formula expressing

the orthogonal invariant $\mathfrak{I}_s^n(G)$ in terms of the μ_k's.
Such a formula reads:

$$\mathfrak{I}_s^n(G) = \sum_{h_1 < \ldots < h_s} \mu_{h_1}^n \ldots \mu_{h_s}^n.$$

9. The Orthogonal Invariants Method for Computing Lower Bounds to the Eigenvalues of Problem (1.2.1)

We come now to answer question (2), posed in the former Section. It is helpful, we believe, to provide an exposition of some considerations of analytical geometry, which are exactly those leading to formula (1.9.7) of this Section, which answers the above-mentioned question (2).

Consider in the Cartesian space X^r ($r \geqslant 3$) an ellipsoid E, centred on the origin of X^r. Let

$$\sum_{ij}^{1,r} g_{ij} x_i x_j = 1 \tag{1.9.1}$$

be its equation with respect to the natural (i.e. rectangular) coordinates of X^r. The left-hand side of (1.9.1) is a positive definite quadratic form in the r real variables $x_1, \ldots, x_r$. Denote by $\alpha_1 \leqslant \ldots \leqslant \alpha_r$ the lengths of the r half-axes of E, arranged in non-decreasing order. If $\mu_1 \geqslant \ldots \geqslant \mu_r$ are the eigenvalues of the above-mentioned quadratic form, one has, as is known, $\alpha_k^2 = \mu_k^{-1}$.

Let us intersect E by a linear space S_ν, passing through the centre of E, of dimension ν with $1 < \nu < r$. Let E_ν be the ellipsoid resulting from this intersection between E and S_ν (Fig. 1.4).

Let $\tilde{\alpha}_1 \leqslant \ldots \leqslant \tilde{\alpha}_\nu$ be the lengths of the ν half-axes of E_ν. One has, as is known,

$$\alpha_k \leqslant \tilde{\alpha}_k \qquad (k = 1, \ldots, \nu),$$

so that, putting $\tilde{\mu}_k = \tilde{\alpha}_k^{-2}$, we will have $\tilde{\mu}_k \leqslant \mu_k$.

Consider the 'orthogonal invariants' I_s of E and the orthogonal invariants $\tilde{I}_s$ of E_ν. As is known from analytical geometry, one has:

$$I_0 = 1, \qquad I_s = \sum_{h_1 < \ldots < h_s}^{1,r} \mu_{h_1} \ldots \mu_{h_s} \qquad (s = 1, \ldots, r), \tag{1.9.2}$$

$$\tilde{I}_0 = 1, \qquad \tilde{I}_s = \sum_{h_1 < \ldots < h_s}^{1,\nu} \tilde{\mu}_{h_1} \ldots \tilde{\mu}_{h_s} \qquad (s = 1, \ldots, \nu). \tag{1.9.3}$$

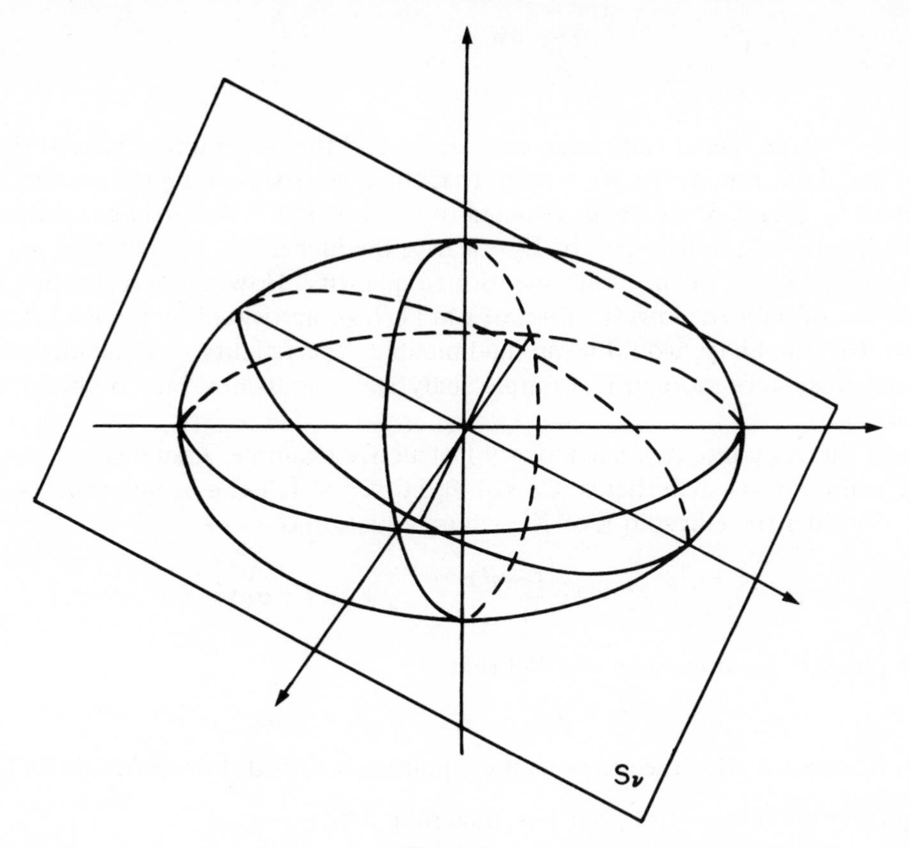

Fig. 1.4.

Let k be an index chosen among $1, \ldots, \nu$. Consider the $\nu - 1$-dimensional linear space $S^k_{\nu-1}$ contained in S_ν and orthogonal to the half-axis of E_ν, whose length has been indicated by $\tilde{\alpha}_k$. The ellipsoid $E^k_{\nu-1}$, intersection of $S^k_{\nu-1}$ with E, will have half-axes of lengths $\tilde{\alpha}_1 \leqslant \ldots \leqslant \tilde{\alpha}_{k-1} \leqslant \tilde{\alpha}_{k+1} \leqslant \ldots \leqslant \tilde{\alpha}_\nu$.

It follows that the orthogonal invariants $I^*_{s,k}$ of $E^k_{\nu-1}$ will be given by

$$I^*_{0,k} = 1, \qquad I^*_{s,k} = \sum_{h_1 < \ldots < h_s}^{1,\nu\;(k)} \tilde{\mu}_{h_1} \ldots \tilde{\mu}_{h_s} \qquad (s = 1, \ldots, \nu - 1), \tag{1.9.4}$$

where the symbol $\sum_{h_1 < \ldots < h_s}^{1,\nu\;(k)}$ means that the sum is extended to any combination of class s of the $\nu - 1$ indices $1, \ldots, k-1, k+1, \ldots, \nu$.

Suppose now we want a lower and an upper bound to the eigenvalue μ_k. Suppose, furthermore, that we know the lengths of the half-axes of E_ν (and, hence, the numbers $\tilde{\mu}_1, \ldots, \tilde{\mu}_\nu$) and an orthogonal invariant I_s $(0 \leqslant s \leqslant \nu)$ of E.

One has then

$$\tilde{\mu}_k \leqslant \mu_k \leqslant \frac{I_s - \tilde{I}_s}{I^*_{s-1,k}} + \tilde{\mu}_h \,. \tag{1.9.5}$$

The lower bound had already been considered and the upper one is proved through simple elementary arguments. We remark that the quantity yielding the second bound is expressed by means of elements assumed to be known, i.e. the orthogonal invariant I_s and the lengths of the half-axes of E_ν. Clearly, the higher is s, the more complicated is the analytical aspect of the above-mentioned quantity. However, a restriction to the consideration of only the lowest values of s in (1.9.5), motivated by a misleading endeavour for simplicity, would be an inadmissible superficiality: for the important thing is not to achieve more or less simple analytical expressions, but to obtain the best bounds to μ_k.

Consider, in this context, a particularly instructive example. Assume $r = 5$. Let ρ and σ be two real numbers such that $0 < \rho < 1$ and $0 < \sigma < 1$. Let ϵ be an arbitrary positive number. Consider the ellipsoid E of X^5, whose equation is

$$\frac{1 + \epsilon^2 \sigma^3 \rho}{1 + \epsilon^2} x_1^2 + \frac{\epsilon^2 + \sigma^3 \rho}{1 + \epsilon^2} x_2^2 + 2\epsilon \frac{1 - \sigma^3 \rho}{1 + \epsilon^2} x_1 x_2 + \rho x_3^2 + \sigma \rho x_4^2 + \sigma^2 \rho x_5^2 = 1.$$

In the present case it is easily verified that

$$\mu_1 = 1, \qquad \mu_2 = \rho, \qquad \mu_3 = \sigma\rho, \qquad \mu_4 = \sigma^2 \rho, \qquad \mu_5 = \sigma^3 \rho.$$

Let us intersect E with the space whose equation is $x_2 = 0$. For the ellipsoid E_4, resulting from the intersection, one has, assuming $\rho < \dfrac{1}{1 + \epsilon^2}$:

$$\tilde{\mu}_1 = \frac{1 + \epsilon^2 \sigma^3 \rho}{1 + \epsilon^2}, \qquad \tilde{\mu}_2 = \rho, \qquad \tilde{\mu}_3 = \sigma\rho, \qquad \tilde{\mu}_4 = \sigma^2 \rho.$$

Set

$$\eta_s = \frac{I_s - \tilde{I}_s}{I^*_{s-1,4}} \qquad s = 1, 2, 3, 4.$$

After some tedious but elementary computations it is found that

$$\eta_1 = \frac{\epsilon^2}{1 + \epsilon^2} + \mathcal{O}(\sigma^3)\,\rho,$$

$$\eta_2 = \epsilon^2 \rho + \mathcal{O}(\sigma)\,\rho + \mathcal{O}(\sigma^3)\,\rho^2,$$

$$\eta_3 = \epsilon^2 \sigma\rho + \mathcal{O}(\sigma^2)\,\rho + \mathcal{O}(\sigma^4)\,\rho^2,$$

$$\eta_4 = \epsilon^2 \sigma^2 \rho + \mathcal{O}(\sigma^3)\,\rho + \mathcal{O}(\sigma^5)\,\rho^2.$$

It follows that ρ and σ may be chosen in such a way that:

$$\eta_1 > \eta_2 > \eta_3 > \eta_4.$$

Therefore, in the particular case under consideration, the higher is s the better is the upper bound to μ_4 given by (1.9.5).

Formula (1.9.5) can be extended from the finite dimensional case to that relative to an 'infinite dimensional' space, i.e. to the Hilbert space S. Consider to this end problem (1.2.2). In this case again we can consider the 'ellipsoid' E of the space S, whose 'equation' is $(Gu, u) = 1$ and again we can 'intersect' such an ellipsoid with a ν-dimensional linear manifold W_ν of S. Let $w_1, w_2, \ldots, w_\nu$ be ν vectors spanning W_ν. It is easily seen that the lengths of the half-axes of the resulting ellipsoid are the inverse square roots of the solutions of the following secular equation:

$$\det\{(Gw_i, w_j) - \mu\,(w_i, w_j)\} = 0 \qquad (i, j = 1, \ldots, \nu). \tag{1.9.6}$$

Let these roots be: $\mu_1^{(\nu)} \geqslant \ldots \geqslant \mu_k^{(\nu)} \geqslant \ldots \geqslant \mu_\nu^{(\nu)}$. Such numbers, as it is easy to prove, are none other than the (lower, in this case) approximations given by the Rayleigh–Ritz method.

In other words, intersecting E by a ν-dimensional linear manifold is equivalent, as far as the computation of the half-axes lengths is concerned, to none other than the application of the Rayleigh–Ritz method. Thereafter it is easy to extend (1.9.5). It will be necessary, assuming $G \in \mathcal{C}^n$, to consider the ellipsoid $(G^n u, u) = 1$ instead of that introduced above ($n = 1$). Let P_ν be the orthogonal projection from S onto W_ν. The only positive eigenvalues of the operator $P_\nu G P_\nu$ are the numbers $\mu_1^{(\nu)} \geqslant \ldots \geqslant \mu_\nu^{(\nu)}$. Denoting by $w_k^{(\nu)}$ an eigenvector of the operator $P_\nu G P_\nu$ corresponding to the eigenvalue $\mu_k^{(\nu)}$, let $W_{\nu,k}$ be the totality of the vectors of W_ν orthogonal to $w_k^{(\nu)}$. Let $P_{\nu,k}$ be the orthogonal projector of S onto $W_{\nu,k}$. Setting, for $s > 0$,

$$\sigma_k^{(\nu)} = \left\{ \frac{\mathfrak{I}_s^n(G) - \mathfrak{I}_s^n(P_\nu G P_\nu)}{\mathfrak{I}_{s-1}^n(P_{\nu,k} G P_{\nu,k})} + [\mu_k^{(\nu)}]^n \right\}^{1/n},^{(8)} \tag{1.9.7}$$

it is proved ([26], [32]), using the results of the theory recalled in Section 8, that

$$\sigma_k^{(\nu)} \geqslant \sigma_k^{(\nu+1)}, \qquad \lim_{\nu\to\infty} \sigma_k^{(\nu)} = \mu_k.$$

Therefore, assuming $\tau_k^{(\nu)} = [\sigma_k^{(\nu)}]^{-1}$, the lower bounds sequence to λ_k, required by problem (1), is constructed.

As far as the computation of $\sigma_k^{(\nu)}$ is concerned, the following considerations are in order. The quantities $\mathfrak{I}_s^n(P_\nu G P_\nu)$, $\mathfrak{I}_{s-1}^n(P_{\nu,k} G P_{\nu,k})$, $\mu_k^{(\nu)}$ are known, once the roots of (1.9.6) have been obtained, i.e. once the Rayleigh–Ritz approximations have been computed. For, $\mathfrak{I}_s^n(P_\nu G P_\nu)$ and $\mathfrak{I}_{s-1}^n(P_{\nu,k} G P_{\nu,k})$ are computed by means of formulae analogous to (1.9.3) and (1.9.4) (with $\tilde{\mu}_{h_i}$ replaced by $[\mu_{h_i}^{(\nu)}]^n$). Therefore one has to know only the orthogonal invariant $\mathfrak{I}_s^n(G)$ for some s and some n (such that $\mathfrak{I}_s^n(G)$ is finite). This shows the interest of the result obtained, which leads the computation of an orthogonal invariants sequence of G: $\mu_1, \mu_2, \ldots, \mu_k, \ldots$ to the computation of a single orthogonal invariant of G. From the theoretical standpoint $\mathfrak{I}_s^n(G)$ is known

---

(8) To simplify the symbols, we shall not indicate the dependence of $\sigma_k^{(\nu)}$ on s and n.

explicitly, since it is given by (1.8.1). It must be kept in mind, however, that, in order to obtain an upper bound $\sigma_k^{(\nu)}$ to μ_k, it will be necessary to know an upper bound to $\mathfrak{I}_s^n(G)$, whereas (1.8.1) only gives lower bounds to such a quantity. Therefore, so far the method can be applied only to those operators G, for which the remainder of the series (1.8.1) can be majorized. There exist several important examples where this is possible but, if the theory could not proceed any further, there would be, as far as the application of the method is concerned, a theoretical restriction not unlike the one relative to the Weinstein–Aronszajn method, which consists in assuming the explicit knowledge of a base problem.

We shall, on the contrary, show in the following Sections how, for some classes of operators of fundamental importance in applications (such as the integral operators and the elliptic differential operators), the problem of computing $\mathfrak{I}_s^n(G)$ can be completely solved.

10. The Orthogonal Invariants of Integral Operators

Let us represent the Hilbert space S by a space $\mathfrak{L}^2(A, \mu)$, A being a subset of a given abstract set (in particular, a Cartesian space), where a non-negative measure μ has been introduced (in particular, the Lebesgue measure). Of course A will be assumed μ-measurable.(9) Let $G \subset \mathfrak{C}^n$. It is then possible to prove ([26], [32]) that G^n is an integral operator

$$G^n u = \int_A K(x, y)\, u(y)\, d\mu_y \tag{1.10.1}$$

K (x, y) being of the following type:

$$K(x, y) = \int_A H(x, t)\, H(t, y)\, d\mu_t$$

where H (x, y) is an Hermitian kernel (i.e. $H(x, y) = \overline{H(y, x)}$) belonging to $\mathfrak{L}^2(A \times A, \mu \times \mu)$. Now, if the integral representation (1.10.1) of G^n is known, i.e. if the kernel K (x, y) is known, all the orthogonal invariants $\mathfrak{I}_s^n(G)$ of G can be constructed. For, consider the function f, depending on the s points $x^{(1)}, \ldots, x^{(s)}$, defined in the following way:

$$f(x^{(1)}, \ldots, x^{(s)}) = \det\{K(x^{(i)}, x^{(j)})\} \qquad (i, j = 1, \ldots, s).$$

This function may be proved to belong to $\mathfrak{L}^1(\underset{1}{A} \times \ldots \times \underset{s}{A}, \underset{1}{\mu} \times \ldots \times \underset{s}{\mu})$.

The following integral representation for $\mathfrak{I}_s^n(G)$ holds:

---

(9) It is convenient to let μ be arbitrary and not to restrict our considerations to subsets of X^r with the Lebesgue measure, in order to give a unified treatment for eigenvalue problems relative to single equations as well as to systems of equations (*see* [26], [32]).

$$\mathfrak{I}_s^n(\mathrm{G}) = \frac{1}{s!} \int_{\mathrm{A}} \cdots \int_{\mathrm{A}} f(x^{(1)}, \ldots, x^{(s)})\, d\mu_{x^{(1)}} \cdots d\mu_{x^{(s)}}. \tag{1.10.2}$$

The proof may be found in [26] and [32].

Inserting this expression of $\mathfrak{I}_s^n(\mathrm{G})$ into (1.9.7), the term $\sigma_k^{(\nu)}$ of the upper approximations sequence to μ_k is obtained by means of 'quadratures' and of a solution of a secular equation (1.9.6).

Problem (1) of page 11 is thus completely solved, whenever the integral representation (1.10.1) is known. Hence, as a particular case, the problem is solved as far as Fredholm integral equations are concerned. It must be mentioned that the formula, obtainable by (1.9.7) through insertion of (1.10.2), was already discovered by Trefftz [82], for the particular case $s = 1$, $n = 2$. Trefftz remarks, however, the comparative uselessness of the above-mentioned formula in eigenvalue problems relative to differential operators (i.e. problems of type (1.2.1)) because, except for some particular cases, the integral representation of the inverse operator (or of an iterated one of the inverse operator) is generally not known; i.e., in other words, the 'Green function' of the problem is unknown. We shall later show how this basic difficulty may be overcome for elliptic operators. This is one of the most delicate points of the theory. Before closing this Section, let us make two remarks. In the first place, the integral representation (1.10.2) for $\mathfrak{I}_s^n(\mathrm{G})$ was in some sense to be expected. For, introducing the Fredholm *entire transcendent function* $\delta(\lambda; \mathrm{K})$, relative to the kernel $\mathrm{K}(x, y)$, it can be noticed – as is known – that it is represented by the following power series:

$$\delta(\lambda; \mathrm{K}) = 1 + \sum_{s=1}^{\infty} (-1)^s \delta_s \lambda^s,$$

where δ_s stands for the right-hand side of (1.10.2). Now, experts of integral equation theory know that $\delta(\lambda; \mathrm{K})$ may be regarded as the 'characteristic function' of the integral operator (1.10.1); therefore, according to what happens for an Hermitian matrix, the coefficients of $\delta(\lambda; \mathrm{K})$ were to be expected to coincide exactly with the orthogonal invariants of the operator G^n.

11. Eigenvalue Problems for Ordinary Differential Equations

The theory, developed in Sections 9 and 10 above, allows us to perform the computation of the eigenvalues for the boundary value problems, where the Green function is explicitly known or, at least, a kernel, allowing the translation of the problem into an integral equation, is known. This is possible for a number of boundary value problems concerning ordinary differential equations. Several problems, studied by the writer and his coworkers, will now be reviewed.

(a)

$$\frac{d^2u}{dx^2} + \lambda (1 + \sin x) u = 0 \qquad (1.11.1)$$

$$u(0) = u(\pi) = 0.$$

The investigation of such a particularly simple problem was motivated by the fact that it has been already considered by Collatz [4], who obtained the following bounds for the first eigenvalue

$$0.5388 < \lambda_1 < 0.54088$$

and by Weinstein [84], who gave the following bounds for the first two eigenvalues:

$$0.5394 < \lambda_1 < 0.54088$$

$$2.35775 < \lambda_2 < 2.38228.$$

Problem (1.11.1) can be translated into the following integral equation:

$$\varphi(x) = \lambda \int_0^{\pi} H(x, y)\, \varphi(y)\, dy \qquad (1.11.2)$$

where

$$H(x, y) = (1 + \sin x)^{1/2}\, G(x, y)\, (1 + \sin y)^{1/2},$$

$$G(x, y) \begin{cases} = y\left(1 - \frac{x}{\pi}\right) & \text{for } 0 \leqslant y \leqslant x \leqslant \pi, \\ = x\left(1 - \frac{y}{\pi}\right) & \text{for } 0 \leqslant x \leqslant y \leqslant \pi. \end{cases}$$

Putting $x = \frac{\pi}{2} + t,\ y = \frac{\pi}{2} + \tau$,

$$K(t, \tau) = H\left(\frac{\pi}{2} + t, \frac{\pi}{2} + \tau\right),$$

$$K_0(t, \tau) = \frac{1}{4}\ \{K(t, \tau) + K(t, -\tau) + K(-t, \tau) + K(-t, -\tau)\},$$

$$K_1(t, \tau) = \frac{1}{4}\ \{K(t, \tau) - K(t, -\tau) - K(-t, \tau) + K(-t, -\tau)\}.$$

Problem (1.11.2) separates out into the following two problems:

$$\Phi_0(t) = \lambda \int_{-\pi/2}^{\pi/2} K_0(t, \tau)\, \Phi_0(\tau)\, d\tau, \qquad \Phi_0(\tau) = \Phi_0(-\tau); \qquad (1.11.3)$$

$$\Phi_1(t) = \lambda \int_{-\pi/2}^{\pi/2} K_1(t, \tau)\, \Phi_1(\tau)\, d\tau, \qquad \Phi_1(\tau) = -\Phi_1(-\tau). \tag{1.11.4}$$

The first problem yields the eigenvalues of (1.11.1) relative to the eigensolutions, belonging to the space S_0 of the functions symmetric with respect to $x = \frac{\pi}{2}$, the second one the eigenvalues relative to the eigensolutions of the space S_1 of the functions anti-symmetric with respect to $x = \frac{\pi}{2}$.

In both cases the application of the Rayleigh–Ritz method has been made computing the zeros of the secular equation:

$$\det \left\{ \int_0^\pi \frac{dv_i}{dx} \frac{dv_j}{dx}\, dx - \lambda \int_0^\pi (1 + \sin x)\, v_i v_j\, dx \right\} = 0 \qquad (i, j = 1, \ldots, \nu)$$

choosing in the two cases, respectively,

$$v_i = \sin (2i - 1)\, x \qquad v_i = \sin 2ix \qquad (i = 1, \ldots, \nu).$$

Using the invariants $\mathfrak{I}_1^2$, both in (1.11.3) and in (1.11.4), and assuming $\nu = 5$, the following bounds to the first ten eigenvalues are obtained [85]:

	Space S_0			Space S_1	
	Lower bounds	Upper Bounds		Lower bounds	Upper bounds
λ_1	0.540282	0.540319	λ_1	2.3708	2.3719
λ_2	5.411	5.449	λ_2	9.70	9.763
λ_3	14.54	15.313	λ_3	21.39	22.10
λ_4	25.29	30.116	λ_4	35.75	39.37
λ_5	34.6	51.8	λ_5	50	61.57

The results relative to this problem (for the case of the space S_0) have been considerably improved through the more sophisticated procedure to be described in Section 13.

(b) *Mathieu's equation*

$$\frac{d^2u}{dx^2} - (8 \cos^2 x)\, u + \lambda u = 0, \tag{1.11.5}$$

$$u'(0) = u'\left(\frac{\pi}{2}\right) = 0.$$

Consider first the more general problem

$$\frac{d^2u}{dx^2} - [p\ (x)]^2 u + \lambda u = 0 \tag{1.11.6}$$

$$u'(0) = u'(l) = 0 \qquad (l > 0.$$

Its corresponding *Rayleigh–Ritz functional* is

$$R\,(u) = \frac{\int_0^l \left(\frac{du}{dx}\right)^2 dx + \int_0^l p^2 u^2\, dx}{\int_0^l u^2\, dx}$$

Let $\{w_k(x)\}$ be a complete orthonormal system in $\mathfrak{L}^2(0, l)$. One has in $\mathfrak{L}^2(0, l)$:

$$p\ (x)\, u\ (x) = \sum_{k=1}^{\infty} w_k(x) \int_0^l p\ (y)\, u\ (y)\, w_k\ (y)\, dy. \tag{1.11.7}$$

Set

$$R_m\,(u) = \frac{\int_0^l \left(\frac{du}{dx}\right)^2 dx + \sum_{k=1}^{m} \left(\int_0^l p u w_k dx\right)^2}{\int_0^l u^2\, dx}.$$

The *Eulerian* eigenvalue problem of the functional $R_m\,(u)$ is the following one:

$$\frac{d^2u}{dx^2} - p\ (x) \sum_{k=1}^{m} w_k(x) \int_0^l p\ (y)\, u\ (y)\, w_k(y)\, dy + \lambda u = 0 \tag{1.11.8}$$

$$u'(0) = u'(l) = 0.$$

Let α be an arbitrary positive number and let v_k be the solution of the problem

$$-\frac{d^2 v_k}{dx^2} + \alpha^2 v_k = p w_k(x), \qquad v_k'(0) = v_k'(l) = 0.$$

One has:

$$v_k(x) = \int_0^l N_\alpha(x, y)\, p\ (y)\, w_k(y)\, dy,$$

where

$$\mathrm{N}_\alpha(x, y) \begin{cases} = \dfrac{\cosh \alpha y \cosh \alpha (l-x)}{\alpha \sinh \alpha l}, & 0 \leqslant y \leqslant x \leqslant l. \\ = \dfrac{\cosh \alpha x \cosh \alpha (l-y)}{\alpha \sinh \alpha l}, & 0 \leqslant x \leqslant y \leqslant l. \end{cases}$$

Putting $\mu = \dfrac{1}{\lambda + \alpha^2}$, problem (1.11.8) is translated into the following integral equation:

$$\int_0^l \mathrm{G}(x, y)\, \varphi(y)\, dy - \mu\varphi(x) = 0, \tag{1.11.9}$$

where

$$\mathrm{G}(x, y) = \mathrm{N}_\alpha(x, y) - \sum_{h,k}^{1,m} a_{hk}\, v_h(x)\, v_k(y)$$

$((a_{hk}))$ $(h, k = 1, \ldots, m)$ being the inverse of the following matrix:

$$\left(\left(\int_0^l \int_0^l \mathrm{N}_\alpha(x, y)\, p(x)\, p(y)\, w_i(x)\, w_j(y)\, dx\, dy + \delta_{ij}\right)\right).$$ [10]

The inequality

$$\mathrm{R}_m(u) \leqslant \mathrm{R}(u)$$

ensures that the eigenvalues of problem (1.11.8) are majorized by the corresponding ones of problem (1.11.6) and that they are, through (1.11.7), as close as one wants to these last ones, provided m is taken large enough.

The orthogonal invariants method can be applied to (1.11.9), since in this case the invariants are very easily computed.

Assuming $l = \frac{\pi}{2}$, $p(x) = 2\sqrt{2} \cos x$, problem (1.11.5) is recovered. Applying the Rayleigh–Ritz method by means of the following system of functions $\{\cos(2k-2)x\}$ $(k = 1, \ldots)$ and using the orthogonal invariant $\mathfrak{J}_1^1$ for (1.11.9), written for $m = 10$, the following bounds are obtained for the eigenvalues of problem (1.11.5) [69], [86].

---

(10) $\delta_{ij} \begin{cases} = 0 & \text{for } i \neq j \\ = 1 & \text{for } i = j. \end{cases}$

	Lower bounds	Upper bounds
λ_1	2.4855	2.4861
λ_2	9.1687	9.1727
λ_3	20.124	20.142
λ_4	39.99	40.06
λ_5	67.85	68.04
λ_6	103.6	104.03
λ_7	147.1	148.02
λ_8	198.5	200.02
λ_9	257.4	260.01
λ_{10}	324	328.01

(c) *Transversal vibrations of a wedge-shaped simply supported beam* [87]

We have to deal here with the eigenvalue problem (1.1.2). The method of Picone, whose results are reported in Section 1 of this survey, has been already applied to the computation of the eigenvalues of this problem.

Tricomi [88] was the first author to obtain bounds on $\sqrt{\lambda_1}$, for the cases $\theta = 0.1, 0.2, 0.4, 0.5, 0.6, 0.7$.

For $\theta = 0.5$ Tricomi's bounds are

$$\pi^2 0.7014 < \sqrt{(\lambda_1)} < \pi^2 0.7660.$$

Bazley and D. W. Fox [39] considered the problem

$$\frac{d^2}{d\xi^2}\left[(1+a\xi)^3 \frac{d^2 u}{d\xi^2}\right] - \sigma (1+a\xi) u = 0 \qquad (1.11.10)$$

$$u(0) = u(\pi) = u''(0) = u''(\pi) = 0,$$

a being a positive constant. Performing the change of variable $x = -\xi\pi^{-1} + 1$ and putting $a = \pi^{-1}\theta(1-\theta)^{-1}$, problem (1.11.10) reduces to problem (1.1.2) and the eigenvalue σ_k $(k = 1, 2, \ldots)$ of (1.11.10) can be expressed through λ_k in the following way:

$$\sigma_k = \frac{\lambda_k}{\pi^4(1-\theta)^2}. \qquad (1.11.11)$$

The following table of lower bounds (obtained through the Weinstein method) and of upper bounds to the first five σ_k for the case $a = \pi^{-1}$, i.e. $\theta = 0.5$ is to be found in [59].

σ_k	Lower bounds	Upper bounds
$k = 1$	2.0826	2.0826
$k = 2$	34.419	34.420
$k = 3$	173.37	173.38
$k = 4$	546.33	546.40
$k = 5$	1331.4	1331.8

Problem (1.1.2) has been deeply studied by Colautti [87] who, among other things applied the orthogonal invariants method.

Putting

$$G(x, \xi; \theta) \begin{cases} = \dfrac{1}{\theta^3}\left\{\dfrac{\theta}{2}\dfrac{\xi x}{1-\theta x} + \dfrac{\theta}{2}\dfrac{\xi x}{1-\theta\xi} + \xi \log(1-\theta x) + x\log(1-\theta\xi)\right. \\ \left. -x\xi\log(1-\theta) - \log(1-\theta\xi) - \dfrac{\theta}{2}\dfrac{\xi}{1-\theta x} - \dfrac{\theta}{2}\dfrac{\xi}{1-\theta\xi}\right\} \text{ for } \xi \leqslant x, \\ = \dfrac{1}{\theta^3}\left\{\dfrac{\theta}{2}\dfrac{x\xi}{1-\theta\xi} + \dfrac{\theta}{2}\dfrac{x\xi}{1-\theta x} + x\log(1-\theta\xi) + \xi\log(1-\theta x)\right. \\ \left. -\xi x\log(1-\theta) - \log(1-\theta x) - \dfrac{\theta}{2}\dfrac{x}{1-\theta\xi} - \dfrac{\theta}{2}\dfrac{x}{1-\theta x}\right\} \text{ for } \xi \geqslant x, \end{cases}$$

let

$$\Gamma(x, \xi; \theta) = (1-\theta x)^{1/2}(1-\theta\xi)^{1/2} G(x, \xi; \theta)$$

and

$$\Gamma\psi = \int_0^1 \Gamma(x, \xi; \theta)\, \psi(\xi)\, d\xi.$$

Problem (1.1.2) is equivalent to the following integral problem:

$$\psi - \lambda\Gamma\psi = 0 \qquad \psi \in \mathfrak{L}^2(0, 1).$$

Two orthogonal invariants of Γ are given by

$$\mathfrak{I}_1^1(\Gamma) = \frac{1}{\theta^2}\left\{\left[-\frac{1}{3\theta^3}+\frac{1}{2\theta^2}-\frac{1}{3\theta}+\frac{1}{12}\right]\log(1-\theta)+\frac{1}{3}\left[-\frac{1}{\theta^2}+\frac{1}{\theta}-\frac{7}{12}\right]\right\},$$

$$\mathfrak{I}_1^2(\Gamma) = \frac{1}{\theta^4}\left\{\left[\frac{1}{9\theta^6}-\frac{1}{3\theta^5}+\frac{17}{36\theta^4}-\frac{7}{18\theta^3}+\frac{7}{36\theta^2}-\frac{1}{18\theta}+\frac{1}{144}\right][\log(1-\theta)]^2\right.$$

$$+\left[\frac{1}{15\theta^5}-\frac{1}{6\theta^4}+\frac{7}{18\theta^3}-\frac{5}{12\theta^2}-\frac{53}{270\theta}-\frac{37}{1080}\right]\log(1-\theta)$$

$$\left.+\left[-\frac{2}{45\theta^4}+\frac{4}{45\theta^3}+\frac{47}{540\theta^2}-\frac{71}{540\theta}+\frac{1427}{32400}\right]\right\}.$$

One has for $\theta = 0.5$:

$$\mathfrak{I}_1^1(\Gamma) = 0.0212914592,$$

$$\mathfrak{I}_1^2(\Gamma) = 0.0003902682363.$$

The Rayleigh–Ritz method is applied using the following functions:

$$\alpha_n(x) = 6x^{n+4} - (n+3)(n+4)x^3 + (n^2+7n+6)x \qquad (n = 1, 2, \ldots).$$

The eigenvalues of (1.1.2), which are not majorized through the Rayleigh–Ritz method, are majorized using the second one of the following inequalities:

$$(1-\theta)^3\pi^4k^4 \leqslant \lambda_k \leqslant \frac{\pi^4k^4}{1-\theta}.$$

These are in turn obtained as a consequence of the inequalities satisfied by the Rayleigh–Ritz functional relative to (1.1.2)

$$(1-\theta)^3\frac{\int_0^1 (u'')^2\,dx}{\int_0^1 u^2\,dx} \leqslant \frac{\int_0^1 (1-\theta x)^3(u'')^2\,dx}{\int_0^1 (1-\theta x)u^2\,dx} \leqslant \frac{1}{1-\theta}\frac{\int_0^1 (u'')^2\,dx}{\int_0^1 u^2\,dx}.$$

Using $\mathfrak{I}_1^2(\Gamma)$ one obtains, for the case $\theta = 0.5$, the following bounds on λ_k and, hence, the bounds on σ_k of the second table, given by (1.11.11), together with those of the third table on $\pi^{-2}\sqrt{(\lambda_k)}$.

λ_k	Lower bounds	Upper bounds
$k = 1$	50.71623063	50.71623066
$k = 2$	838.2089	838.2091
$k = 3$	4222.235	4222.247
$k = 4$	13305.47	13305.82
$k = 5$	32427	32432
$k = 6$	67140	67185
$k = 7$	124108	124390
$k = 8$	210726	212111
$k = 9$	334060	339657
$k = 10$	498394	517576
$k = 11$	701261	757963
$k = 12$	928773	1074337
$k = 13$	1166861	1504837

σ_k	Lower bounds	Upper bounds
$k = 1$	2.082607695	2.082607697
$k = 2$	34.420153	34.420157
$k = 3$	173.3815	173.3821
$k = 4$	546.37	546.39
$k = 5$	1331.57	1331.79
$k = 6$	2757	2759
$k = 7$	5096	5108
$k = 8$	8653	8711
$k = 9$	13717	13948
$k = 10$	204660	212537
$k = 11$	287965	311250
$k = 12$	381390	441165
$k = 13$	479159	617995

$\pi^{-2}\sqrt{\lambda_k}$	Lower bounds	Upper bounds
$k = 1$	0.7215621412	0.7215621415
$k = 2$	2.9334345	2.9334348
$k = 3$	6.583722	6.583731
$k = 4$	11.6873	11.6875
$k = 5$	18.245	18.247
$k = 6$	26.253	26.263
$k = 7$	35.69	35.74
$k = 8$	46.51	46.67
$k = 9$	58.59	59.06
$k = 10$	71.52	72.90
$k = 11$	84.84	88.22
$k = 12$	97.64	105.02
$k = 13$	109.4	124.3

(d) *Differential equation concerning a heat conduction problem*

$$\frac{\mathrm{d}}{\mathrm{d}x}\left[(1-x^7)\frac{\mathrm{d}u}{\mathrm{d}x}\right] + \lambda x^7 u = 0 \tag{1.11.12}$$

$$u(0) = 0, \qquad u(x) \text{ continuous at } x = 1.$$

To this difficult eigenvalue problem, formerly studied by several authors [89], [90], [91], [92], [93], [94], with rather limited success, has been applied in [95], [96], the orthogonal invariants method.

The difficulty of the problem is due to the fact that one has to deal with a differential equation, which is singular at $x = 1$; so that no quantitative condition at $x = 1$ is imposed on the unknown function u, but only a continuity condition at such a point.

Setting:

$$\gamma(x) = \int_0^x \frac{d\xi}{1-\xi^7}; \qquad G(x, \xi) \begin{cases} = \gamma(\xi) & 0 \leqslant \xi \leqslant x < 1, \\ = \gamma(x) & 0 \leqslant x \leqslant \xi < 1, \end{cases}$$

$$k(x, \xi) = x^{7/2} G(x, \xi)\, \xi^{7/2},$$

problem (1.11.12) is equivalent to the following integral problem:

$$w(x) = \lambda \int_0^1 k(x, \xi)\, w(\xi)\, d\xi$$

so that the orthogonal invariants method can be applied.

The justification of the Rayleigh–Ritz method is rather delicate, because of the singular nature of the differential operator under examination. This justification, however, is completely achieved in [95]. Using the invariant ϑ_1^2, the following bounds are obtained [96]:

$$8.721575 < \lambda_1 < 8.727471$$

$$128.2512 < \lambda_2 < 152.4231$$

$$208.3475 < \lambda_3 < 435.0634.$$

In the following table we report the results formerly obtained by other authors:

	λ_1	λ_2	λ_3	λ_4	λ_5
Latzko (1921)	8.712	164.36	1700.40		
Durfee (1956)	8.72747	152.423	435.06	855.68	1414.1
Fettis (1957)	8.72798	152.8	462.5		
Caligo (1962)	$8.7275 < \lambda_1 < 8.7448$				
Caligo–Cotugno (1964)	$8.72752 < \lambda_1 < 8.72759$				

It is proved in [95] that the lower bounds obtained by Caligo [93] and by Caligo and Cotugno [94] are misleading.

In [97] the problem of the asymptotic distribution of the eigenvalues of (1.11.12) is considered. Let us remember that, in this case, the known procedures cannot be applied, because the differential operator under consideration is a singular one. Denoting, as usual, by λ the number of all those eigenvalues not exceeding $N(\lambda)$ one has:

$$N(\lambda) \cong \left(\frac{1}{\pi} \int_0^1 \sqrt{\frac{x^7}{1-x^7}}\, dx \right) \sqrt{\lambda}$$

and hence

$$\lambda_n \cong \left[\frac{\pi}{\int_0^1 \sqrt{\frac{x^7}{1-x^7}}\, dx} \right]^2 n^2.$$

(e) *Computation of eigenvalues connected with a genetics problem*

$$\frac{d^2 u}{dx^2} + 2\alpha \frac{d}{dx} [(x - x_0) u] + \frac{4\lambda}{1-x^2} u = 0$$

$$u(-1) = u(1) = 0.$$

One has to compute the eigenvalues λ of this problem for several choices of the parameters α and x_0.

The translation of the problem into an integral equation and the application of the Rayleigh–Ritz method, together with the orthogonal invariants one, produced the following results [98]:

$\alpha = 0$

	Lower bounds	Upper bounds	Exact value
λ_1	0.49999	0.50001	0.5
λ_2	1.49999	1.50001	1.5
λ_3	2.99999	3.00001	3
λ_4	4.99999	5.00001	5
λ_5	7.49999	7.50001	7.5
λ_6	10.49999	10.50001	10.5
λ_7	13.99999	14.00001	14
λ_8	17.99999	18.00001	18
λ_9	22.49999	22.50001	22.5
λ_{10}	27.49999	27.50001	27.5

$\alpha = 0.4 \quad x_0 = 0$

	Lower bounds	Upper bounds
λ_1	0.4239	0.4244
λ_2	1.446	1.451
λ_3	2.93	2.953
λ_4	4.9	4.96
λ_5	7.34	7.46
λ_6	10.24	10.46
λ_7	13.58	13.96
λ_8	17.35	17.96
λ_9	21.51	22.46
λ_{10}	26.06	27.46

$\alpha = 0.4 \quad x_0 = 0.4$

	Lower bounds	Upper bounds
λ_1	0.429	0.4295
λ_2	1.45	1.4541
λ_3	2.939	2.956
λ_4	4.91	4.96
λ_5	7.35	7.46
λ_6	10.25	10.46
λ_7	13.59	13.96
λ_8	17.36	17.96
λ_9	21.53	22.46
λ_{10}	26.08	27.46

$\alpha = 0.4 \quad x_0 = 0.6$

	Lower bounds	Upper bounds
λ_1	0.4353	0.4358
λ_2	1.4544	1.4587
λ_3	2.94	2.961
λ_4	4.91	4.97
λ_5	7.35	7.47
λ_6	10.24	10.47
λ_7	13.58	13.97
λ_8	17.34	17.97
λ_9	21.49	22.47
λ_{10}	26.03	27.47

$\alpha = 0.4 \quad x_0 = 1$

	Lower bounds	Upper bounds
λ_1	0.4554	0.4561
λ_2	1.4677	1.4736
λ_3	2.95	2.971
λ_4	4.91	4.98
λ_5	7.32	7.48
λ_6	10.19	10.48
λ_7	13.47	13.98
λ_8	17.15	17.98
λ_9	21.20	22.48
λ_{10}	25.60	27.48

$\alpha = 1 \quad x_0 = 0$

	Lower bounds	Upper bounds
λ_1	0.3262	0.3266
λ_2	1.398	1.404
λ_3	2.88	2.91
λ_4	4.83	4.91
λ_5	7.25	7.41
λ_6	10.11	10.41
λ_7	13.39	13.91
λ_8	17.06	17.91
λ_9	21.1	22.41
λ_{10}	25.49	27.41

$\alpha = 1 \quad x_0 = 0.4$

	Lower bounds	Upper bounds
λ_1	0.356	0.3566
λ_2	1.4209	1.4284
λ_3	2.89	2.91
λ_4	4.83	4.93
λ_5	7.23	7.43
λ_6	10.04	10.43
λ_7	13.25	13.93
λ_8	16.82	17.93
λ_9	20.7	22.43
λ_{10}	24.9	27.43

$\alpha = 1 \quad x_0 = 0.6$	Lower bounds	Upper bounds
λ_1	0.3931	0.3939
λ_2	1.4504	1.4593
λ_3	2.91	2.96
λ_4	4.85	4.96
λ_5	7.22	7.46
λ_6	10.01	10.46
λ_7	13.18	13.96
λ_8	16.7	17.96
λ_9	20.5	22.46
λ_{10}	24.6	27.46

$\alpha = 1 \quad x_0 = 1$	Lower bounds	Upper bounds
λ_1	0.5106	0.5122
λ_2	1.5444	1.5584
λ_3	2.985	3.038
λ_4	4.89	5.035
λ_5	7.22	7.54
λ_6	9.93	10.54
λ_7	12.98	14.04
λ_8	16.3	18.04
λ_9	19.9	22.54
λ_{10}	23.7	27.54

$\alpha = 2 \quad x_0 = 0$	Lower bounds	Upper bounds
λ_1	0.2016	0.20193
λ_2	1.386	1.3989
λ_3	2.83	2.89
λ_4	4.72	4.88
λ_5	7.03	7.38
λ_6	9.7	10.38
λ_7	12.7	13.88
λ_8	15.9	17.88
λ_9	19.4	22.38
λ_{10}	23.1	27.38

$\alpha = 2 \quad x_0 = 0.4$	Lower bounds	Upper bounds
λ_1	0.2982	0.2992
λ_2	1.49	1.513
λ_3	2.89	2.98
λ_4	4.73	4.97
λ_5	6.96	7.46
λ_6	9.51	10.46
λ_7	12.3	13.96
λ_8	15.3	17.96
λ_9	18.5	22.46
λ_{10}	21.7	27.46

$\alpha = 2 \quad x_0 = 0.6$	Lower bounds	Upper bounds
λ_1	0.416	0.418
λ_2	1.624	1.655
λ_3	2.98	3.09
λ_4	4.79	5.07
λ_5	6.97	7.57
λ_6	9.45	10.57
λ_7	12.1	14.06
λ_8	15	18.07
λ_9	18	22.56
λ_{10}	21.1	27.57

$\alpha = 2 \quad x_0 = 1$	Lower bounds	Upper bounds
λ_1	0.7702	0.7794
λ_2	2.049	2.116
λ_3	3.27	3.44
λ_4	4.99	5.41
λ_5	7.05	7.9
λ_6	9.34	10.9
λ_7	11.8	14.39
λ_8	14.3	18.4
λ_9	16.9	22.9
λ_{10}	19.6	27.9

12. The Orthogonal Invariants of Second Order Ordinary Linear Differential Operators

Recently Colautti [99] has considered the eigenvalue problem

$$Eu + \lambda u \equiv \frac{d}{dx}\left[\theta(x)\frac{du}{dx}\right] - c(x)u + \lambda u = 0$$

$$u(0) = u(1) = 0 \tag{1.12.1}$$

$$(\theta(x) > 0, \qquad c(x) \geqslant 0),$$

with the aim of estimating the orthogonal invariants of the corresponding Green operator Γ, without explicitly constructing the Green function $G(x, \xi)$ or making use of special devices, as in some examples considered in the former Section.

Consider the boundary value problem $Eu + f = 0$, $u(0) = u(1) = 0$. Denoting its Green function by $G(x, \xi)$, i.e. expressing the solution as

$$u(x) = \int_0^1 G(x, \xi) f(\xi)\, d\xi \equiv \Gamma f,$$

we wish to compute the three orthogonal invariants $\mathfrak{J}_1^1(\Gamma)$, $\mathfrak{J}_1^2(\Gamma)$, $\mathfrak{J}_1^3(\Gamma)$.

Let $v_1(x)$ and $v_2(x)$ be the integrals of the equation $Ev = 0$ satisfying the conditions:

$$v_1(1) = 0, \qquad v_1'(1) = -1$$

$$v_2(0) = 0, \qquad v_2'(0) = 1.$$

The following equalities are proved in [99]:

$$\mathfrak{J}_1^1(\Gamma) = \frac{1}{\theta(1) v_2(1)} \int_0^1 v_1(x)\, v_2(x)\, dx$$

$$\mathfrak{J}_1^2(\Gamma) = \frac{2}{[\theta(1) v_2(1)]^2} \int_0^1 [v_1(x)]^2\, dx \int_0^x [v_2(y)]^2\, dy$$

$$\mathfrak{J}_1^3(\Gamma) = \frac{6}{[\theta(1) v_2(1)]^3} \int_0^1 [v_1(x)]^2\, dx \int_0^x v_1(y) v_2(y)\, dy \int_0^y [v_2(z)]^2\, dz.$$

Let $f_1(x)$, $f_2(x)$ be two arbitrary functions, ϵ_1 and ϵ_2 two positive numbers such that

$$|v_1(x) - f_1(x)| < \epsilon_1, \qquad |v_2(x) - f_2(x)| < \epsilon_2 \qquad (0 \leqslant x \leqslant 1)$$

and suppose $f_2(1) > \epsilon_2$.

Set

$$Q_1 = \frac{1}{\theta(1) f_2(1)} \int_0^1 f_1(x) f_2(x)\, dx$$

$$Q_2 = \frac{2}{[\theta(1) f_2(1)]^2} \int_0^1 [f_1(x)]^2\, dx \int_0^x [f_2(y)]^2\, dy$$

$$Q_3 = \frac{6}{[\theta(1) f_2(1)]^3} \int_0^1 [f_1(x)]^2\, dx \int_0^x f_1(y) f_2(y)\, dy \int_0^y [f_2(z)]^2\, dz.$$

Since in applications to eigenvalue theory an upper bound to $\mathfrak{J}_1^n(\Gamma)$ is needed, let ρ_1, ρ_2, ρ_3 be three positive numbers, vanishing with ϵ_1 and ϵ_2, such that

$$\mathfrak{J}_1^m(\Gamma) < Q_m + \rho_m \qquad (m = 1, 2, 3).$$

Since Q_m has to be considered known, $f_1(x)$ and $f_2(x)$ being known, an estimate of ρ_1, ρ_2, ρ_3 is needed.

The following expressions for ρ_1, ρ_2, ρ_3 can be found in [99]:

$$\rho_1 = \frac{1}{\theta(1)[f_2(1)-\epsilon_2]}\left\{\epsilon_1 \int_0^1 f_2(x)\,dx + \epsilon_2 \int_0^1 f_1(x)\,dx + \frac{\epsilon_2}{f_2(1)} \int_0^1 f_1(x) f_2(x)\,dx + \epsilon_1\epsilon_2\right\}.$$

$$\rho_2 = \frac{2}{\{\theta(1)[f_2(1)-\epsilon_2]\}^2}\left\{\frac{\epsilon_2(2f_2(1)-\epsilon_2)}{[f_2(1)]^2} \int_0^1 [f_1(x)]^2\,dx \int_0^x [f_2(y)]^2\,dy\right.$$

$$+ 2\epsilon_1 \int_0^1 f_1(x)\,dx \int_0^x [f_2(y)]^2\,dy + \epsilon_1^2 \int_0^1 dx \int_0^x [f_2(y)]^2\,dy$$

$$+ 2\epsilon_2 \int_0^1 [f_1(x)]^2\,dx \int_0^x f_2(y)\,dy + 4\epsilon_1\epsilon_2 \int_0^1 f_1(x)\,dx \int_0^x f_2(y)\,dy$$

$$\left. + 2\epsilon_1^2\epsilon_2 \int_0^1 dx \int_0^x f_2(y)\,dy + \epsilon_2^2 \int_0^1 x[f_1(x)]^2\,dx + 2\epsilon_1\epsilon_2^2 \int_0^1 x f_1(x)\,dx + \frac{\epsilon_1^2\epsilon_2^2}{2}\right\}.$$

$$\rho_3 = \frac{6}{[\theta(1)(f_2(1)-\epsilon_2)]^3}\left\{\left(\frac{3\epsilon_2}{f_2(1)} - \frac{3\epsilon_2^2}{[f_2(1)]^2} + \frac{\epsilon_2^3}{[f_2(1)]^3}\right).\right.$$

$$\cdot \int_0^1 [f_1(x)]^2\,dx \int_0^x f_1(y) f_2(y)\,dy \int_0^y [f_2(z)]^2\,dz$$

$$+ 2\epsilon_2 \int_0^1 [f_1(x)]^2\,dx \int_0^x f_1(y) f_2(y)\,dy \int_0^y f_2(z)\,dz + \epsilon_2^2 \int_0^1 [f_1(x)]^2\,dx \int_0^x y f_1(y) f_2(y)\,dy$$

$$+\epsilon_1 \int_0^1 [f_1(x)]^2\,dx \int_0^x f_2(y)\,dy \int_0^y [f_2(z)]^2\,dz + \epsilon_2 \int_0^1 [f_1(x)]^2\,dx \int_0^x f_1(y)\,dy \int_0^y [f_2(z)]^2\,dx$$

$$+\epsilon_1\epsilon_2 \int_0^1 [f_1(x)]^2\,dx \int_0^x dy \int_0^y [f_2(z)]^2\,dz + 2\epsilon_1\epsilon_2 \int_0^1 [f_1(x)]^2\,dx \int_0^x f_2(y)\,dy \int_0^y f_2(z)\,dz$$

$$+2\epsilon_2^2 \int_0^1 [f_1(x)]^2\,dx \int_0^x f_1(y)\,dy \int_0^y f_2(z)\,dz + 2\epsilon_1\epsilon_2^2 \int_0^1 [f_1(x)]^2\,dx \int_0^x dy \int_0^y f_2(z)\,dz$$

$$+\epsilon_1\epsilon_2^2 \int_0^1 [f_1(x)]^2\,dx \int_0^y y f_2(y)\,dy + \epsilon_2^3 \int_0^1 [f_1(x)]^2\,dx \int_0^x y f_1(y)\,dy$$

$$+\frac{\epsilon_1\epsilon_2^3}{2} \int_0^1 x^2[f_1(x)]^2\,dx + 2\epsilon_1 \int_0^1 f_1(x)\,dx \int_0^x f_1(y) f_2(y)\,dy \int_0^y [f_2(z)]^2\,dz$$

$$+\epsilon_1^2 \int_0^1 dx \int_0^x f_1(y) f_2(y)\,dy \int_0^y [f_2(z)]^2\,dz + 4\epsilon_1\epsilon_2 \int_0^1 f_1(x)\,dx \int_0^x f_1(y) f_2(y)\,dy \int_0^y f_2(z)\,dz$$

$$+2\epsilon_1^2\epsilon_2 \int_0^1 dx \int_0^x f_1(y) f_2(y)\,dy \int_0^y f_2(z)\,dz + 2\epsilon_1\epsilon_2^2 \int_0^1 f_1(x)\,dx \int_0^x y f_1(y) f_2(y)\,dy$$

$$+\epsilon_1^2\epsilon_2^2 \int_0^1 dx \int_0^x y f_1(y) f_2(y)\,dy + 2\epsilon_1^2 \int_0^1 f_1(x)\,dx \int_0^x f_2(y)\,dy \int_0^y [f_2(z)]^2\,dz$$

$$+2\epsilon_1\epsilon_2 \int_0^1 f_1(x)\,dx \int_0^x f_1(y)\,dy \int_0^y [f_2(z)]^2\,dz + 2\epsilon_1^2\epsilon_2 \int_0^1 f_1(x)\,dx \int_0^x dy \int_0^y [f_2(z)]^2\,dz$$

$$+\epsilon_1^3 \int_0^1 dx \int_0^x f_2(y)\,dy \int_0^y [f_2(z)]^2\,dz + \epsilon_1^2\epsilon_2 \int_0^1 dx \int_0^x f_1(y)\,dy \int_0^y [f_2(z)]^2\,dz$$

$$+\epsilon_1^3\epsilon_2 \int_0^1 dx \int_0^x dy \int_0^y [f_2(z)]^2\,dz + 4\epsilon_1^2\epsilon_2 \int_0^1 f_1(x)\,dx \int_0^x f_2(y)\,dy \int_0^y f_2(z)\,dz$$

$$+2\,\epsilon_1^3\epsilon_2\int_0^1 dx\int_0^x f_2(y)\,dy\int_0^y f_2(z)\,dz+2\,\epsilon_1^2\epsilon_2^2\int_0^1 f_1(x)\,dx\int_0^x yf_2(y)\,dy$$

$$+\epsilon_1^3\epsilon_2^2\int_0^1 dx\int_0^x yf_2(y)\,dy+4\,\epsilon_1\epsilon_2^2\int_0^1 f_1(x)\,dx\int_0^x f_1(y)\,dy\int_0^y f_2(z)\,dz$$

$$+2\,\epsilon_1^2\epsilon_2^2\int_0^1 dx\int_0^x f_1(y)\,dy\int_0^y f_2(z)\,dz+2\,\epsilon_1\epsilon_2^3\int_0^1 f_1(x)\,dx\int_0^x yf_1(y)\,dy$$

$$+\epsilon_1^2\epsilon_2^3\int_0^1 dx\int_0^x yf_1(y)\,dy+4\,\epsilon_1^2\epsilon_2^2\int_0^1 f_1(x)\,dx\int_0^x dy\int_0^y f_2(z)\,dz$$

$$+2\,\epsilon_1^3\epsilon_2^2\int_0^1 dx\int_0^x dy\int_0^y f_2(z)\,dz+\epsilon_1^2\epsilon_2^3\int_0^1 x^2f_1(x)\,dx+\frac{\epsilon_1^3\epsilon_2^3}{6}\Bigg\}.$$

Now an estimate of ϵ_1, ϵ_2, ϵ_3 is needed. To this end, it is proved in [99] that, if one has

$$Eu=f,\qquad u(0)=a,\qquad u'(0)=b,$$

the following estimate holds:

$$\max_{[0,1]}|u(x)|\leqslant e^{\mathrm{M}}\max_{[0,1]}\left|a+b\theta(0)\int_0^x\frac{d\xi}{\theta(\xi)}+\int_0^x f(t)\left(\int_t^x\frac{d\xi}{\theta(\xi)}\right)dt\right| \tag{1.12.2}$$

where

$$\mathrm{M}=\max_{[0,1]}|c(x)|\int_0^1\frac{d\xi}{\theta(\xi)}. \tag{1.12.3}$$

If, on the contrary,

$$Eu=f,\qquad u(1)=a,\qquad u'(1)=b,$$

one has

$$\max_{[0,1]}|u(x)|\leqslant e^{\mathrm{M}}\max_{[0,1]}\left|a+b\theta(1)\int_1^x\frac{d\xi}{\theta(\xi)}+\int_1^x f(t)\left(\int_t^x\frac{d\xi}{\theta(\xi)}\right)dt\right|, \tag{1.12.4}$$

M being given by (1.12.3).

Applying (1.12.4) to the function $u = v_1 - f_1$ and (1.12.2) to $u = v_2 - f_2$, the following expressions for ϵ_1 and ϵ_2 are obtained:

$$\epsilon_1 = e^{\mathrm{M}} \max_{[0,1]} \left| f_1(1) + (1 + f_1'(1))\,\theta(1) \int_1^x \frac{d\xi}{\theta(\xi)} + \int_1^x \mathrm{E}f_1 \left(\int_t^x \frac{d\xi}{\theta(\xi)} \right) dt \right|$$

$$\epsilon_2 = e^{\mathrm{M}} \max_{[0,1]} \left| f_2(0) + (f_2'(0) - 1)\,\theta(0) \int_0^x \frac{d\xi}{\theta(\xi)} + \int_0^x \mathrm{E}f_2 \left(\int_t^x \frac{d\xi}{\theta(\xi)} \right) dt \right| .$$

Finally, to complete the research, Colautti proves that, denoting by $\{p_k(x)\}$ the monomial system or the classical trigonometric one, the constants $c_{1k}^{(n)}, c_{2k}^{(n)}$ $(k = 1, \ldots, n)$ may be determined in such a way that, setting

$$f_1^{(n)}(x) = \sum_{k=1}^{n} c_{1k}^{(n)} p_k(x), \qquad f_2^{(n)}(x) = \sum_{k=1}^{n} c_{2k}^{(n)} p_k(x)$$

and considering the $\mathrm{Q}_m^{(n)}$'s and $\rho_m^{(n)}$'s corresponding to such functions $(m = 1, 2, 3)$, one has:

$$\lim_{n \to \infty} (\mathrm{Q}_m^{(n)} + \rho_m^{(n)}) = \mathfrak{J}_1^m(\Gamma) \qquad (m = 1, 2, 3).$$

To this purpose, it is enough to take as $c_{1k}^{(n)}$ and $c_{2k}^{(n)}$, respectively, the solutions of the following systems:

$$\sum_{k=1}^{n} \left[\int_0^1 \mathrm{E}p_h\, \mathrm{E}p_k\, dx + p_h(1)\, p_k(1) + p_h'(1)\, p_k'(1) \right] c_{1k}^{(n)} = - p_h'(1)$$

$$\sum_{k=1}^{n} \left[\int_0^1 \mathrm{E}p_h\, \mathrm{E}p_k\, dx + p_h(0)\, p_k(0) + p_h'(0)\, p_k'(0) \right] c_{2k}^{(n)} = p_h'(0)$$

$(h = 1, \ldots, n)$.

Another procedure for computing $f_1^{(n)}$ and $f_2^{(n)}$, suggested in [99], comes from the approximate values of v_1 and v_2 obtained through the successive approximations method applied to the differential problems, whose solutions are v_1 and v_2. Such a method is more cumbersome to apply, but it has the advantage of exhibiting the order of magnitude of the errors ϵ_1 and ϵ_2, relative to the n-th approximation.

More precisely one has:

$$| v_i(x) - f_i^{(n)}(x) | = \mathcal{O}\left(\frac{|K|^n}{(n+3)!}\right) \qquad (i = 1, 2),$$

K being a constant depending on $\theta(x)$ and $c(x)$.(11)

13. Improvement of the Lower Bounds Using a Computer of Limited Capacity

Let us consider again the general eigenvalue problem (1.2.1), within the hypotheses stated in Section 2.

If the orthogonal invariants method, using the invariant $\mathfrak{I}_1^2$, is applied in order to get lower bounds to the λ_k's and if the Rayleigh–Ritz one is applied to get upper bounds, we have, indicating by $\lambda_1^{(n)}, \ldots, \lambda_n^{(n)}$ the n-th approximation given by the Rayleigh–Ritz method, $\lambda_k \leqslant \lambda_k^{(n)}$ and, furthermore,

$$\lambda_k > \left\{ \frac{1}{(\lambda_k^{(n)})^2} + \mathfrak{I}_1^2(G) - \sum_{h=1}^{n} \frac{1}{(\lambda_h^{(n)})^2} \right\}^{-1/2}. \tag{1.13.1}$$

The lower bound given by (1.13.1) tends to λ_k as $n \to \infty$, as shown by the theory recalled in Section 9. Hence, in order to improve the lower approximation, n has to be increased more and more, i.e..the upper approximation must be improved. Now this argument is purely theoretical. In fact, given an arbitrary electronic computer, there corresponds some fixed n_0 (the larger the more powerful the computer is) such that it is impossible to compute, by means of the given computer, the characteristic roots of a matrix of rank $m > n_0$ (as required by the Rayleigh–Ritz method) to a degree of accuracy fixed in advance. It follows that the eigenvalue computation (upper and lower bounds) cannot be pushed beyond the n_0-th approximation. Now the numerical experiments show that, as far as the upper approximations are concerned, i.e. the values given by the Rayleigh–Ritz method, the present electronic computers are able to yield satisfactory approximations, i.e. their relative n_0 is large enough to meet the requirements of applications. This is not the case, to the same degree, for the lower approximations, especially as far as higher order eigenvalues are concerned. In other words, if the Rayleigh–Ritz method gives satisfactory approximations for the first k eigenvalues, the lower approximations to be considered as satisfactory are, in general, relative to the first k' eigenvalues, with $k' < k$. The repeated numerical experiments, performed by the writer and his coworkers have always confirmed the occurrence of this circumstance.

The following problem then arises: *for a fixed n_0, assuming there to be only the possibility of computing eigenvalues of $n \times n$ matrices, with $n \leqslant n_0$, to improve the lower bounds given, for instance, by* (1.13.1).

This is a theoretical problem, investigated in the two Notes [71], by means of rather sophisticated functional analysis techniques. The results obtained consist in the following two theorems, which will be stated assigning to the symbols that occur the same meaning as in Sections 2 and 9 of this survey.

(11) Cassisa [184] has extended Colautti's results to the case of a general boundary value problem for the equation $Eu + \lambda u = 0$.

1.IX. *Let* $G \in \mathfrak{C}^2$. *Let* $v_1, \ldots, v_n$ *be n linearly independent vectors of* V. *Let* $z_1, \ldots, z_q$ *be q* ($q \geqslant 2$) *linearly independent vectors of* V, *such that* $Lz_r \in V$ ($r = 1, \ldots, q$). *One has*:

$$\lambda_k > \left\{ \frac{1}{(\lambda_k^{(n)})^2} + \vartheta_1^2(G) - \sum_{h=1}^{n} \frac{1}{(\lambda_h^{(n)})^2} - \sum_{i=2}^{q} \nu_i^{(q)} \right\}^{-1/2} \tag{1.13.2}$$

$\lambda_1^{(n)} \leqslant \ldots \leqslant \lambda_n^{(n)}$ *being the roots of the secular equations*

$$\det \{(Lv_i, v_j) - \lambda(v_i, v_j)\} = 0 \qquad (i, j = 1, \ldots, n) \tag{1.13.3}$$

and $\nu_1^{(q)} \geqslant \nu_2^{(q)} \geqslant \ldots \geqslant \nu_q^{(q)}$ *the q roots (all non-negative) of the secular equation*

$$\det \left\{ (z_r, Lz_s) - \sum_{i,k}^{1,n} \beta_{ik} (Lz_r, v_i)(v_k, Lz_s) - \nu (Lz_r, L^2 z_s) \right\} = 0. \tag{1.13.4}$$

$\{\beta_{ik}\}$ *stands for the* $n \times n$ *inverse matrix of the matrix* $\{(Lv_i, v_j)\}$ ($i, j = 1, \ldots, n$).

(1.13.2) yields the required improvement of (1.13.1). It must be kept in mind that $z_1, \ldots, z_q$ can be chosen in such a way that all roots of (1.13.4) are positive.

1.X. *Let* $G \in \mathfrak{C}^2$. *Let* $v_1, \ldots, v_n$ *be n linearly independent vectors of* V *and* $z_1, \ldots, z_q$ *q* ($q \geqslant 2$) *linearly independent vectors of* V. *One has*:

$$\lambda_k > \left\{ \frac{1}{(\tilde{\lambda}_k^{(n)})^2} + \vartheta_1^2(G) - \sum_{h=1}^{n} \frac{1}{(\tilde{\lambda}_h^{(n)})^2} - \sum_{i=2}^{q} \tilde{\nu}_i^q \right\}^{-1/2}, \tag{1.13.5}$$

$\tilde{\lambda}_1^{(n)} \leqslant \ldots \leqslant \tilde{\lambda}_n^{(n)}$ *being the roots of the secular equation*

$$\det \{(Lv_i, Lv_j) - \tilde{\lambda}(v_i, Lv_j)\} = 0 \qquad (i, j = 1, \ldots, n) \tag{1.13.6}$$

and $\tilde{\nu}_1^{(q)} \geqslant \ldots \geqslant \tilde{\nu}_q^{(q)}$ *the q roots (all non-negative) of the secular equation*

$$\det \left\{ (z_r, z_s) - \sum_{i,k}^{1,n} \gamma_{ik} (z_r, Lv_i)(Lv_k, z_s) - \tilde{\nu} (Lz_r, Lz_s) \right\} = 0.$$

$\{\gamma_{ik}\}$ *stands for the* $n \times n$ *inverse matrix of* $\{(Lv_i, Lv_j)\}$ ($i, j = 1, \ldots, n$).

(1.13.5) may not improve (1.13.1), but it improves instead only the bounds obtained through (1.13.1), replacing therein the $\lambda_k^{(n)}$'s by the $\tilde{\lambda}_k^{(n)}$'s. Since the $\tilde{\lambda}_k^{(n)}$'s yield upper bounds to the λ_k's that are no better than the $\lambda_k^{(n)}$'s, a worse lower bound is obtained as a result of replacing $\lambda_k^{(n)}$ by $\tilde{\lambda}_k^{(n)}$ in (1.13.1). This last bound is in turn improved by (1.13.5), which, however, in practice improves also (1.13.1) and, sometimes, (1.13.2) itself.

On the other hand, (1.13.5) has the great advantage, with respect to (1.13.2), of employing vectors z_r, which are *not* such that they require Lz_k to belong to V, as, on the contrary, is required by (1.13.2).

We present here the results obtained by applying the *improved formulae* given by Theorems 1.IX and 1.X to the problem considered in Section 11, Example (a), for the space S_0 only.

The following tables show the various approximations for n, ranging between 2 and 15. In each table there are four columns after the one indicating the eigenvalues to be approximated. In the first column are reported the lower bounds given by the 'not improved' formula (1.13.1) and in the last one the Rayleigh–Ritz upper bounds obtained solving (1.13.3). In the second column there are the lower bounds given by the *improved formula* (1.13.2) and in the third one the lower bounds obtained through (1.13.5).

Near the lower bounds given by the *improved formulae*, there are some integer numbers in brackets: each of them stand for the value needed for the order n of the approximation, in order to achieve a lower bound, *through the not improved formula* (1.13.1), not worse than that indicated by the number near the integer in brackets. In this way the advantage of the *improved formulae* can be observed.For instance, they yield already for 10-th approximation lower bounds, which formula (1.13.1) can yield, *not worse*, only to the 14-th or to the 15-th approximation.

$n = 2$

	(1.13.1)	(1.13.2)		(1.13.5)		(1.13.3)
λ_1	0.5397	0.5398	(3)	0.5397	(2)	0.54033
λ_2	5	5.07	(3)	4.97	(2)	5.497

$n = 3$

	(1.13.1)	(1.13.2)		(1.13.5)		(1.13.3)
λ_1	0.54015	0.54019	(4)	0.54018	(4)	0.54032
λ_2	5.28	5.32	(4)	5.3	(4)	5.449
λ_3	12.6	13.3	(4)	13.05	(4)	15.6

$n = 4$

	(1.13.1)	(1.13.2)		(1.13.5)		(1.13.3)
λ_1	0.54024	0.54027	(5)	0.54026	(5)	0.5403189
λ_2	5.377	5.402	(5)	5.395	(5)	5.4487
λ_3	13.92	14.37	(5)	14.22	(5)	15.314
λ_4	22.7	24.8	(5)	24.4	(5)	30.97

$n = 5$

	(1.13.1)	(1.13.2)		(1.13.5)		(1.13.3)
λ_1	0.540282	0.540297	(6)	0.540296	(6)	0.54031887
λ_2	5.411	5.427	(6)	5.424	(6)	5.44865
λ_3	14.54	14.85	(6)	14.79	(6)	15.3129
λ_4	25.29	27.04	(7)	26.68	(6)	30.1156
λ_5	34.6	39.5	(7)	38.9	(7)	51.8

$n = 6$

	(1.13.1)	(1.13.2)		(1.13.5)		(1.13.3)
λ_1	0.540297	0.5403	(7)	0.5403	(7)	0.540318862
λ_2	5.427	5.437	(8)	5.436	(8)	5.448639
λ_3	14.85	15.06	(8)	15.03	(8)	15.3127
λ_4	27.02	28.36	(8)	28.18	(8)	30.1156
λ_5	38.6	42.9	(8)	42.3	(8)	49.86
λ_6	48	57	(8)	56	(8)	78.2

$n = 7$

	(1.13.1)	(1.13.2)		(1.13.5)		(1.13.3)
λ_1	0.540305	0.540312	(9)	0.540312	(9)	0.540318861
λ_2	5.435	5.442	(9)	5.441	(9)	5.4486368
λ_3	15.01	15.16	(9)	15.15	(9)	15.31263
λ_4	28.04	29.05	(9)	28.95	(9)	30.1152
λ_5	41.8	45.4	(9)	45	(8)	49.855
λ_6	53	61	(9)	60	(9)	74.55
λ_7	63	78	(10)	77	(10)	111

$n = 8$

	(1.13.1)	(1.13.2)		(1.13.5)		(1.13.3)
λ_1	0.540309	0.540314	(10)	0.540314	(10)	0.5403188599
λ_2	5.439	5.444	(11)	5.444	(11)	5.4486364
λ_3	15.11	15.22	(11)	15.21	(11)	15.31262
λ_4	28.68	29.43	(11)	29.38	(11)	30.1151
λ_5	44.05	46.94	(11)	46.72	(11)	49.8536
λ_6	58.4	65.7	(11)	65.1	(11)	74.53
λ_7	69	83	(11)	82	(11)	104.3
λ_8	79	101	(11)	100	(11)	148

$n = 9$

	(1.13.1)	(1.13.2)		(1.13.5)		(1.13.3)
λ_1	0.540312	0.540316	(12)	0.540316	(12)	0.5403188597
λ_2	5.442	5.445	(11)	5.445	(11)	5.4486363
λ_3	15.17	15.25	(12)	15.24	(12)	15.312611
λ_4	29.08	29.66	(12)	29.62	(12)	30.11501
λ_5	45.56	47.87	(12)	47.73	(12)	49.8534
λ_6	62.08	68.36	(12)	67.95	(12)	74.528
λ_7	76	89	(12)	88	(12)	104.14
λ_8	87	107	(12)	106	(12)	138.9
λ_9	96	128	(13)	126	(13)	192

$n = 10$

	(1.13.1)	(1.13.2)		(1.13.5)		(1.13.3)
λ_1	0.540314	0.540317	(14)	0.5403169	(14)	0.54031885962
λ_2	5.443	5.446	(13)	5.446	(13)	5.4486362
λ_3	15.20	15.27	(14)	15.26	(13)	15.31261
λ_4	29.35	29.79	(14)	29.77	(14)	30.115
λ_5	46.61	48.46	(14)	48.37	(14)	49.85331
λ_6	64.83	70.1	(14)	69.82	(14)	74.527
λ_7	81.6	92.9	(14)	92.3	(14)	104.137
λ_8	95	115	(14)	113	(14)	138.7
λ_9	105	135	(14)	133	(14)	179
λ_{10}	115	157	(15)	155	(14)	243

$n = 11$

	(1.13.1)	(1.13.2)		(1.13.5)		(1.13.3)
λ_1	0.540315	0.5403175	(15)	0.5403174	(15)	0.54031885959
λ_2	5.445	5.447	(15)	5.447	(15)	5.44863618
λ_3	15.23	15.28	(15)	15.28	(15)	15.312609
λ_4	29.53	29.88	(15)	29.87	(15)	30.11499
λ_5	47.35	48.84	(> 15)	48.78	(15)	49.85329
λ_6	66.88	71.28	(> 15)	71.08	(15)	74.5269
λ_7	85.8	95.78	(> 15)	95.32	(15)	104.136
λ_8	102	120	(15)	119	(15)	138.68
λ_9	115	143	(15)	142	(15)	178.2
λ_{10}	125	164	(> 15)	162	(15)	224
λ_{11}	135	189	(> 15)	187	(> 15)	299

$n = 12$

	(1.13.1)	(1.13.2)	(1.13.5)	(1.13.3)
λ_1	0.540316	0.5403178	0.5403178	0.54031885958
λ_2	5.4459	5.4476	5.4475	5.44863617
λ_3	15.25	15.29	15.28	15.3126087
λ_4	29.66	29.94	29.93	30.114988
λ_5	47.89	49.1	49.06	49.85327
λ_6	68.43	72.09	71.95	74.52682
λ_7	89.1	97.78	97.44	104.1355
λ_8	108	124	123	138.6793
λ_9	124	150	149	178.16
λ_{10}	136	175	173	223
λ_{11}	146	197	194	274
λ_{12}	155	223	221	362

$n = 13$

	(1.13.1)	(1.13.2)	(1.13.5)	(1.13.3)
λ_1	0.5403167	0.5403181	0.540318	0.54031885957
λ_2	5.4465	5.4478	5.4478	5.44863616
λ_3	15.26	15.295	15.294	15.3126085
λ_4	29.76	29.98	29.98	30.114986
λ_5	48.29	49.28	49.25	49.853262
λ_6	69.60	72.66	72.56	74.52679
λ_7	91.83	99.23	98.98	104.1354
λ_8	112	127	127	138.679
λ_9	131	156	155	178.158
λ_{10}	146	184	182	222.6
λ_{11}	158	209	206	272
λ_{12}	167	231	229	329
λ_{13}	177	260	257	431

$n = 14$

	(1.13.1)	(1.13.2)	(1.13.5)	(1.13.3)
λ_1	0.5403171	0.5403182	0.5403182	0.540318859564
λ_2	5.4469	5.448	5.448	5.448636154
λ_3	15.27	15.299	15.299	15.31260842
λ_4	29.83	30.01	30.01	30.1149854
λ_5	48.59	49.41	49.39	49.853259
λ_6	70.51	73.07	73	74.52678
λ_7	93.94	100.2	100.1	104.13533
λ_8	116	129	129	138.6789
λ_9	137	160	159	178.1574
λ_{10}	155	191	189	222.58
λ_{11}	169	219	217	271.92
λ_{12}	181	245	242	327
λ_{13}	190	269	266	390
λ_{14}	199	299	296	506

$n = 15$

	(1.13.1)	(1.13.2)	(1.13.5)	(1.13.3)
λ_1	0.5403175	0.5403184	0.5403184	0.540318859562
λ_2	5.4472	5.4481	5.4481	5.448636152
λ_3	15.28	15.3	15.3	15.31260838
λ_4	29.88	30.03	30.03	30.114985
λ_5	48.82	49.49	49.48	49.853257
λ_6	71.21	73.34	73.32	74.52677
λ_7	95.62	100.9	100.9	104.1353
λ_8	120	131	131	138.6788
λ_9	143	163	163	178.1572
λ_{10}	163	195	195	222.571
λ_{11}	180	227	226	271.919
λ_{12}	194	255	255	326.21
λ_{13}	204	281	281	386
λ_{14}	213	305	305	456
λ_{15}	223	338	338	588

14. Robert's Formulae for the Orthogonal Invariants

The orthogonal invariants theory has been recently re-investigated, in its theoretical aspect, by the French mathematician Didier Robert [100], who studied its properties of algebraic character in the framework of the Grothendieck algebraic theory of compact operators [101]. The most remarkable aspect of the Robert investigation consists essentially in the discovery of the following two formulae:

$$\mathfrak{J}_s^n(G) = \frac{1}{s}\sum_{q=1}^{s}(-1)^{q-1}\,\mathfrak{J}_1^{qn}(G)\,\mathfrak{J}_{s-q}^n(G);$$

$$\mathfrak{J}_s^n(G) = (-1)^s \sum_{1\leqslant k\leqslant s}\frac{(-1)^k}{k!}\left\{\sum_{\substack{r_1+\ldots+r_k=s\\ r_i\geqslant 1}} \frac{\mathfrak{J}_1^{nr_1}(G)\ldots\mathfrak{J}_1^{nr_k}(G)}{r_1\ldots r_k}\right\}.$$

The interest of Robert's formulae lies in the fact that they relate the computation of the orthogonal invariant $\mathfrak{J}_s^n(G)$, of *degree n* and *order s*, to orthogonal invariants of lower order. Plainly, the second formula relates the computation of $\mathfrak{J}_s^n(G)$ to the computation of first order orthogonal invariants. This is important because, while it is not difficult to compute numerically first order orthogonal invariants also of high degree it is, on the contrary, almost impossible to achieve the numerical computation of the orthogonal invariants of order $n \geqslant 3$, through definition (1.8.1) or (1.10.2).

15. Invariant Subspaces and Ordering of the Bounds

Let us consider, now, another problem of a theoretical nature, but of extreme practical interest, which arises in eigenvalue approximation theory. This problem, which looks simple at first glance, is, on the contrary, far from being trivial. Within the hypotheses and notations of Section 2, assume the Hilbert space S decomposable as a (finite or denumerably infinite) *direct sum* of the spaces $H_1, H_2, \ldots, H_s \ldots$

$$S = H_1 \oplus H_2 \oplus \ldots \oplus H_s \oplus \ldots$$

By this we mean that, for any $u \in S$, one has, in one and only one way,

$$u = u_1 + u_2 + \ldots + u_s + \ldots, \qquad u_s \in H_s$$

(the series on the right-hand side being convergent in the metric of S, if the H_s's are infinite). Let the spaces H_s be pairwise orthogonal and let, furthermore, any H_s be an *invariant subspace* of the operator G, i.e.

$$G(H_s) \subset H_s.$$

Problem (1.2.2) is equivalent to the following system of eigenvalue problems:

$$Gu - \mu^{(s)}u = 0, \qquad u \in H_s \qquad (s = 1, 2, \ldots).$$

Set $V_s = G(H_s)$. It is easy to verify that

$$V = V_1 \oplus V_2 \oplus \ldots \oplus V_s \oplus \ldots$$

and that problem (1.2.1) is equivalent to the following system of eigenvalue problems:

$$Lv - \lambda^{(s)}v = 0, \qquad v \in V_s, \qquad \lambda^{(s)} = \frac{1}{\mu^{(s)}} \qquad (s = 1, 2, \ldots) \tag{1.15.1}$$

Let $\lambda_1^{(s)} \leqslant \lambda_2^{(s)} \leqslant \ldots \leqslant \lambda_k^{(s)} \leqslant \ldots$ be the eigenvalues of problem (1.15.1). Suppose known the following *table* (t_s) of upper and lower bounds to the first p_s eigenvalues of problem (1.15.1)

$$\delta_1^{(s)} \leqslant \lambda_1^{(s)} \leqslant \epsilon_1^{(s)}$$

. . . .

. . . .

$$\delta_{p_s}^{(s)} \leqslant \lambda_{p_s}^{(s)} \leqslant \epsilon_{p_s}^{(s)}.$$

Assume there to have been obtained the upper bounds $\epsilon_k^{(s)}$ through the Rayleigh–Ritz method. The following problem then arises:

Is it possible to deduce, through the bounds contained in the tables (t_s), *bounds on the k-th eigenvalue* λ_k *of problem* (1.2.1)?

In solving such a problem no assumptions will be made about the method used in obtaining the lower bounds $\delta_k^{(s)}$. It will be only assumed, of course without loss of generality, that $\delta_1^{(s)} \leqslant \delta_2^{(s)} \leqslant \ldots \leqslant \delta_{p_s}^{(s)}$.

To make our assumptions more concrete, we shall further suppose the tables (t_s)

given for $s = 1, \ldots, q$ $(q \geq 1)$ and that, for $s > q$, there is a known constant c_s such that $\lambda_k^{(s)} > c_s$ $(s > q;\ k = 1, 2, \ldots)$, with $\lim_{s\to\infty} c_s = +\infty$ if the number of the H_s's is infinite.

Let $\{\delta_k\}$ and $\{\epsilon_k\}$ be the (finite) sequences obtained collecting all numbers $\delta_k^{(s)}$ and $\epsilon_k^{(s)}$ $(s = 1, \ldots, q;\ k = 1, \ldots, p_s)$, respectively, arranged in non-decreasing order.

The following theorem holds, solving the problem posed before [72]:

1.XI. *Let* $\delta_{p_r}^{(r)}$ *be such that* $\delta_{p_r}^{(r)} \leq \delta_{p_r}^{(s)}$, *for* $s = 1, \ldots, q$. *Let* $c_s \geq \delta_{p_r}^{(r)}$ *for any possible* $s > q$. *Let n be the smallest integer such that* $\delta_n = \delta_{p_r}^{(r)}$. *The following bounds hold on the first n eigenvalues of problem* (1.2.1):

$$\delta_k \leq \lambda_k \leq \epsilon_k \qquad (k = 1, \ldots, n).$$

These are the best bounds obtainable by the given information.

Later on we shall see the interest of this theorem in applications.

16. Computation of the Orthogonal Invariants for Elliptic Differential Operators. Special Theory

We will now deal with the case, of greater practical interest, in which the operator L, considered in Section 2, is an elliptic matrix partial differential operator and the manifold V a subspace of S, determined by suitable 'homogeneous boundary conditions'.

This is the most difficult and interesting aspect of the theory, since it leads to some structure theorems for the operator G, which in the present case is the *Green operator* of the elliptic boundary value problem corresponding to the given eigenvalue problem.

We shall first consider some particular aspects of the theory, which are however such as to cover a large class of applications. The general case will be studied later on (*see* Section 24).

If B is a domain of the Cartesian space X^r, we shall denote by $H_m(B)$ the Hilbert space of the l (complex) components vector functions, having weak derivatives up to the m-th order belonging to $L^2(B)$, equipped with the following scalar product:[(12)]

$$(u, v)_{m,B} = \sum_{0 \leq |s| \leq m} \int_B D^s u\, D^s v\, dx.$$

As a preliminary remark, notice that (1.9.7) may be suitably modified replacing G by an operator $G_\rho \in \mathfrak{C}^n$, such that $G_\rho \geq G$, i.e., one has:

$$\sigma_k^{(\nu,\rho)} = \left\{ \frac{\mathfrak{I}_s^n(G_\rho) - \mathfrak{I}_s^n(P_\nu G P_\nu)}{\mathfrak{I}_{s-1}^n(P_{\nu,k} G P_{\nu,k})} + [\mu_k^{(\nu)}]^n \right\}^{1/n}.$$

(12) If a and b are two l-vectors, their scalar product is denoted by $ab = \sum_i a_i \bar{b}_i$.

Let $G_\rho \geqslant G_{\rho+1}$ and $\lim_{\rho\to\infty} \| G_\rho - G \| = 0$. One has then

$$\sigma_k^{(\nu,\rho)} \leqslant \sigma_k^{(\tilde{\nu},\tilde{\rho})} \qquad \text{if } \nu \geqslant \tilde{\nu},\ \rho \geqslant \tilde{\rho}$$

and furthermore

$$\lim_{\substack{\nu\to\infty\\ \rho\to\infty}} \sigma_k^{(\nu,\rho)} = \mu_k.$$

One has then to consider the determination of an operator sequence G_ρ in such a way that $\mathfrak{J}_s^n(G_\rho)$ can be computed for any element of the sequence.

Now let A be a bounded domain of X^r, which will be assumed *properly regular*, according to a definition given by the writer more than thirty years ago.(13) Consider the following $2m$-th order matrix partial differential operator

$$L(x, D) \equiv \sum_{\substack{0\leqslant |p|\leqslant m\\ 0\leqslant |q|\leqslant m}} D^p a_{pq}(x) D^q,$$

where the $a_{pq}(x)$'s are $l \times l$ complex matrices, supposed, for the sake of simplicity, to be of class C^∞ in the whole of X^r.

Assume that the following conditions are satisfied:

(1) For any $x \in X^r$ let there be satisfied the ellipticity condition

$$\det \sum_{\substack{|p|=m\\ |q|=m}} a_{pq}(x)\, \xi^p \xi^q \neq 0$$

for any real non-zero ξ.

(2) $L(x, D)$ is formally self-adjoint, i.e.

$$a_{pq}(x) = (-1)^{|p|+|q|} \overline{a}_{qp}(x)$$

(if a is a matrix, the matrix obtained by transposing its elements and replacing them by the corresponding complex conjugates is denoted by $\overline{a}$).

(3) The 'bilinear form' associated with $L(x, D)$:

$$B(u, v) = \sum_{|p|,|q|}^{0,m} (-1)^{|p|} \int_A (a_{pq} D^q u) D^p v \, dx$$

is such that, for any vector function $u \in C^\infty(X^r)$, one has

$$(-1)^m B(u, u) \geqslant c \sum_{|p|=m} \int_A | D^p u |^2 \, dx,$$

(13) Briefly, A is properly regular if its boundary is *piecewise regular* and satisfies a *cone hypothesis*; see [106] and [32].

c being a positive constant independent of u.

The eigenvalue problem we shall consider is:

$$\begin{cases} Lv - (-1)^m \lambda v = 0 & \text{in A}^{(14)} \\ D^p v = 0 \quad (0 \leqslant |p| \leqslant m-1 & \text{on } \partial A. \end{cases}$$

The boundary conditions, now under consideration, are of the Dirichlet type. In what follows the theory will be extended to more general boundary conditions.

The assumptions made on $L(x, D)$ allow the existence of a linear operator R with the following properties:

(a) R is a continuous operator, with domain $\mathfrak{L}^2(A)$ and range in the Hilbert space $H_{2m}(A)$;

(b) R, viewed as an operator from $\mathfrak{L}^2(A)$ into $\mathfrak{L}^2(A)$, is Hermitian (i.e. symmetric);

(c) for any $f \in \mathfrak{L}^2(A)$, one has $LRf = f$;

(d) E is an integral operator:

$$Rf = \int_A f(y)\, F(x, y)\, dy \tag{1.16.1}$$

and the matrix $F(x, y)$ is a *principal fundamental solution* (in the sense of Giraud), relative to the operator $L(x, D)$ in the domain A.

Consider the space of the (l-complex components) vector functions of class C^∞ in $\bar{A}$, concisely denoted by $C^\infty(\bar{A})$. Let Γ be the (finite dimensional) manifold of all functions such that $B(w, w) = 0$. Denote by $\mathcal{H}(A)$ the Hilbert space obtained through functional completion of the quotient space $C^\infty(\bar{A})/\Gamma$, by means of the norm introduced by the following scalar product:

$$((u, v)) = (-1)^m B(u, v).$$

Denote by $(\cdot\,,\cdot)$ the scalar product in the space $\mathfrak{L}^2(A)$ and let R^* be the linear and continuous operator, with domain in $\mathcal{H}(A)$ and range in $\mathfrak{L}^2(A)$ defined by the condition

$$((Rf, g)) = (f, R^*g)$$

for any $f \in \mathfrak{L}^2(A)$ and any $g \in \mathcal{H}(A)$.

Finally, let P be the orthogonal projector of $\mathcal{H}(A)$ onto its subspace $\Omega(A)$, formed by the solutions of the homogeneous equation $Lu = 0$.

If we denote by $\mathfrak{A}(A)$ the class of the functions belonging to $H_m(A) \cap H_{2m}(B)$ for any B such that $\bar{B} \subset A$, then, for any given $f \in \mathfrak{L}^2(A)$, one proves ([32], [102]), the existence and uniqueness within the class $\mathfrak{A}(A)$ of the solution of the problem

$$\begin{aligned} Lv &= (-1)^m f & &\text{in A,} \\ D^p v &= 0 \quad (0 \leqslant |p| \leqslant m-1) & &\text{on } \partial A. \end{aligned} \tag{1.16.2}$$

Moreover, setting

---

(14) The occurrence of the factor $(-1)^m$ multiplying the parameter λ makes all eigenvalues of the problem positive.

$$G = R^* R - R^* PR,$$

such a solution is given by

$$v = Gf,$$

i.e. G is the *Green operator* of problem (1.16.2).

Let $\{\omega_k\}$ be a complete system of linearly independent vectors in $\Omega(A)$ and let $\Omega_\rho(A)$ be the ρ-dimensional manifold spanned by $\omega_1, \ldots, \omega_\rho$. Define P_ρ as the orthogonal projector of $\mathcal{H}(A)$ onto $\Omega_\rho(A)$. Let us set

$$G_\rho = R^* R - R^* P_\rho R.$$

The operators G and G_ρ belong to $\mathfrak{C}^n$ for any $n > r/2m$. Moreover, one has $G_\rho > G_{\rho+1}$ and $\lim_{\rho\to\infty} \| G - G_\rho \| = 0$.

On the other hand, as soon as the integral representation (1.16.1) for R is known, an integral representation for both $R^* R$ and $R^* P_\rho R$ is easily constructed, so that the computation of the orthogonal invariants of G_ρ is provided by (1.10.2). Let us also remark that, in many cases, it is more convenient, from the numerical standpoint, in order to compute $\mathfrak{I}^n_s(G_\rho)$, to use (1.8.1) instead of (1.10.2), since the particular structure of the operator G_ρ may allow the majorization of the remainder of the series in the right-hand side of (1.8.1).

The theory considered in this Section is a special one, because with the operator L have been associated the particular boundary conditions of Dirichlet type; moreover, the *explicit knowledge* of the operator R, i.e. of a principal fundamental solution, has been assumed.

It is known that a *principal fundamental solution matrix* is explicitly known if the coefficients of the operator are constant [103].

Consider, for instance, the case $a_{pq} \equiv 0$ for $|p| + |q| < 2m$ and a_{pq} constant for $|p| = m$, $|q| = m$, the ellipticity condition

$$\Omega(\xi) \equiv \det \sum_{\substack{|p|=m \\ |q|=m}} a_{pq} \xi^p \xi^q \neq 0$$

being satisfied for any real non-zero ξ.

Let, therefore,

$$L(D) = \sum_{\substack{|p|=m \\ |q|=m}} a_{pq} D^p D^q.$$

Let $\widetilde{L}(\xi)$ be the transpose of the matrix obtained from $L(\xi) = \sum_{\substack{|p|=m \\ |q|=m}} a_{pq} \xi^p \xi^q$ replacing its elements by the corresponding cofactors.

We shall denote by Δ_2 the Laplace operator and, when writing $\Delta_{2(y)}$, we shall mean that it acts on the coordinates $y_1, \ldots, y_r$ of y. Let Ω_ξ be the unit sphere $|\xi| = 1$ in the

r-dimensional cartesian space and let $d\omega_\xi$ be the measure differential on Ω_ξ. If r is odd, set

$$S(x-y) = \frac{1}{4(2\pi i)^{r-1}(2m-1)!}(\Delta_{2(y)})^{(r-1)/2} \int_{\Omega_\xi} \frac{|(x-y)\xi|^{2m-1}}{Q(\xi)} d\omega_\xi$$

and if r is even, set

$$S(x-y) = \frac{1}{(2\pi i)^r(2m)!}(\Delta_{2(y)})^{r/2} \int_{\Omega_\xi} \frac{|(x-y)\xi|^{2m}\log|(x-y)\xi|}{Q(\xi)} d\omega_\xi .$$

The *fundamental solution matrix* is

$$F(x, y) \equiv F(x-y) = \tilde{L}(D)\, S(x-y).$$

17. The Green Matrix and the Eigenvalue Computations in the First Elastostatic Problem

As an illustration of the theory sketched in the former Section, let us consider the case in which the partial differential operator under consideration is the classical one of elastostatics, considered in X^2 or in X^3.

Let us set, therefore,

$$L_i u = \sum_{h=1}^{r} u_{i/hh} + \alpha u_{h/ih} \quad (15)$$

with $r = 2$ or $r = 3$, α standing for a real non-negative constant. Set

$$B(u, v) = -\sum_{i,h}^{1,r} \int_A (u_{i/h} v_{i/h} + \alpha u_{i/i} v_{h/h})\, dx; \qquad \varphi(t) \begin{cases} = \log t^{-1} & \text{for } r = 2 \\ = t^{-1} & \text{for } r = 3. \end{cases}$$

The fundamental matrix $F(x-y)$ relative to the operator $L \equiv (L_1, \ldots, L_r)$, now under consideration, is the classical one of Lord Kelvin [104] and Somigliana [105] and its elements F_{ij} are given by

$$F_{ij}(x-y) = \frac{\alpha}{8\pi(1+\alpha)} \frac{\partial^2 |x-y|^2 \varphi(|x-y|)}{\partial x_i \partial x_j} - \frac{\delta_{ij}}{(r-1)2\pi}\varphi(|x-y|).$$

The operator $Rf \equiv (R_1 f, \ldots, R_r f)$ is defined in the following way:

$$R_i f = \sum_{j=1}^{r} \int_A F_{ij}(x-y) f_j(y)\, dy.$$

---

(15) The symbol $v_{/h}$ stands for $\frac{\partial v}{\partial x_h}$ and hence $v_{/hk} = \frac{\partial^2 v}{\partial x_h \partial x_k}$.

The operator R maps a vector f of $H_0(A)$, that is of $\mathfrak{L}^2(A)$, into a vector of $H_2(A)$ and hence of $\mathfrak{A}(A)$, which in the present case is formed by the r-component vectors belonging to $H_1(A) \cap H_2(B)$ for any B, such that $\overline{B} \subset A$. Moreover one has

$$LRf = f.$$

The domain A, as in the former Section, is assumed bounded and properly regular. Assume further that in the present case the spaces under consideration are formed only by vectors whose components are real.

We want to solve, in the space $\mathfrak{A}(A)$, the boundary problem

$$Lu + f = 0 \qquad \text{in A,} \tag{1.17.1}$$

$$u = 0 \qquad \text{on } \partial A. \tag{1.17.2}$$

The uniqueness of the solution is very easily proved. To show the existence and to construct u in an explicit way, consider Rf as an operator defined in $H_0(A)$ with range in $H_1(A)$. Such an operator is continuous.

Introduce in $H_1(A)$ the scalar product

$$((u, v)) = -B(u, v).$$

This is allowed if we identify two functions of $H_1(A)$ which differ only by an additive constant. Let $\mathcal{H}(A)$ be the space obtained in this way.

Consider the operator R^*, adjoint to R, when R is thought of as an operator mapping $H_0(A)$ into $\mathcal{H}(A)$. In other words, R^* is defined by the condition

$$((Rf, g)) = (f, R^*g)_{0,A}$$

with $f \in H_0(A)$, $g \in \mathcal{H}(A)$. It is not difficult to verify that R^*g has the following representation

$$R_k^*g = \sum_{i,h}^{1,r} \int_A [F_{ik/h}(x-y)\, g_{i/h}(y) + \alpha F_{ik/i}(x-y)\, g_{h/h}(y)]\, dy.$$

We want to find the solution u of problem (1.17.1), (1.17.2) of the following form:

$$u = R^*g.$$

We have to determine g in such a way as to satisfy all conditions of the problem, stated above.

Let A_0 be a domain containing $\overline{A}$ in its interior. Let $\{\varphi_s(x)\}$ $(s = 1, 2, \ldots)$ be a complete system in $\mathfrak{L}^2(A_0 - A)$. Condition (1.17.2) will be satisfied if the function R^*g, considered for $x \in A_0 - A$, is such that

$$\int_{A_0-A} \varphi_s(x)\, R^*v\, dx = 0. \tag{1.17.3}$$

Set

$$\omega_s(y) = \int_{A_0 - A} \varphi_s(x)\, F(x - y)\, dx.$$

The condition may be rewritten in this way

$$((v, \omega_s)) = 0 \qquad (s = 1, 2, \ldots).$$

Let us consider the linear manifold Ω of the solutions of the homogeneous equation $L(D)\,u = 0$ in A, belonging to $\mathcal{H}(A)$. This manifold is closed in $\mathcal{H}(A)$, being closed in $H_1(A)$. Denote by P the orthogonal projector of $\mathcal{H}(A)$ onto Ω. It follows, from what was said above, that the function $u = R^*(v - Pv)$ satisfies (1.17.2).

For $v \in \mathcal{U}(A)$ one has, as may be shown by elementary computations,

$$LR^* v = -Lv.$$

Hence it follows that the function

$$u = R^* Rf - R^* PRf,$$

proved to belong to $\mathcal{U}(A)$, satisfies (1.17.1) and (1.17.2), i.e. is the solution of our problem.

A structure theorem for the Green operator has thus been obtained in the particular case just considered.

Let us point out now, how by means of this theorem and of the orthogonal invariant $\mathfrak{J}_2^1$, one can obtain [26] explicit lower bounds to the eigenvalues of the problem

$$Lu + \lambda u = 0 \quad \text{in} \quad A, \qquad u = 0 \quad \text{on} \quad \partial A.$$

Set

$$\gamma_{ij}(x, y) = -\sum_{h,k}^{1,r} \int_A \{F_{ik/h}(x - t)\, F_{jk/h}(t - y) + \alpha F_{ik/k}(x - t)\, F_{jh/h}(t - y)\}\, dt.$$

Consider a complete system of solutions of the homogeneous equation $Lu = 0$.[16] For the sake of simplicity, suppose the system orthonormalized in the following way:

$$-B(\omega^p, \omega^q) = \delta_{pq}.$$[17]

Set

$$\rho_i^p(x) = \sum_{h,k}^{1,r} \int_A [F_{ik/h}(x - t)\, \omega_{k/h}^p(t) + \alpha F_{ik/k}(x - t)\, \omega_{h/h}^p(t)]\, dt.$$

---

(16) For the construction of such systems *see* [106], ch. III.

(17) Such an orthonormalization is not required in numerical applications and is introduced here only to simplify the formulae.

Let $\{w_i\}$ be a system of functions, vanishing on ∂A, such that $\{Lw_i\}$ is complete in $\mathfrak{L}^2(A)$. Let $\lambda_1^{(\nu)} \leqslant \ldots \leqslant \lambda_\nu^{(\nu)}$ be the solutions of the secular equation

$$\det \{(Lw_i, w_j) + \lambda(w_i, w_j)\} = 0 \qquad (i, j = 1, \ldots, \nu).$$

Putting

$$\tau_k^{(\nu)} = \left\{ \sum_{i,j}^{1,r} \left[\int_A \int_A |\gamma_{ij}(x, y)|^2 \, dx\, dy + \sum_{p,q}^{1,\nu} \int_A \rho_i^p(x) \rho_i^q(x) \, dx \int_A \rho_j^p(x) \rho_j^q(x) \, dx \right.\right.$$

$$\left.\left. - 2 \sum_{p=1}^{\nu} \int_A \int_A \gamma_{ij}(x, y) \rho_i^p(x) \rho_j^p(x) \, dx\, dy \right] - \sum_{i=1}^{\nu}{}^{(k)} \frac{1}{[\lambda_i^{(\nu)}]^2} \right\}^{-1/2} \tag{18}$$

one has:

$$\tau_k^{(\nu)} < \lambda_k \leqslant \lambda_k^{(\nu)}$$

$$\tau_k^{(\nu)} \leqslant \tau_k^{(\nu+1)}, \qquad \lambda_k^{(\nu+1)} \leqslant \lambda_k^{(\nu)}$$

and

$$\lim_{\nu\to\infty} \tau_k^{(\nu)} = \lim_{\nu\to\infty} \lambda_k^{(\nu)} = \lambda_k.$$

18. Iterated Elliptic Operators: Generalized Biharmonic Problem

Let us consider the matrix partial differential operator

$$L(x, D)\, u = \sum_{0 \leqslant |s| \leqslant \nu} a_s(x)\, D^s u,$$

with $x \in X^r$, the $a_s(x)$'s being $l \times l$ complex matrices supposed, for simplicity, to be of class C^∞ in the whole X^r. Let the operator L be elliptic, i.e.

$$\det \sum_{|s|=\nu} a_s(x)\, \xi^s \neq 0$$

for any real ξ.

Consider the operator L^* adjoint to L

$$L^*(x, D)\, u = \sum_{0 \leqslant |s| \leqslant \nu} (-1)^{|s|} D^s [\bar{a}_s(x)\, u].$$

Let A be the usual bounded, properly regular domain of X^r. Let $\mathfrak{U}(A)$ be the space of the (l-complex components) vector functions belonging to $H_\nu(A) \cap H_{2\nu}(B)$ for any

(18) As usual, $\sum_{i=1}^{\nu}{}^{(k)}$ means that the k-th term is not summed.

B, such that $\bar{B} \subset A$. We will consider in $\mathfrak{A}(A)$ the following problem

$$\begin{aligned} &L^* Lu = f && \text{in } A \\ &D^p u = 0 \qquad (0 \leqslant |p| \leqslant \nu - 1) && \text{on } \partial A. \end{aligned} \tag{1.18.1}$$

This problem will be called the *generalized biharmonic problem*. The reason for this nomenclature is that, when $L \equiv \Delta_2$, the classical biharmonic problem is recovered.

Problem (1.18.1) cannot always be considered in the framework of the theory developed in Section 16 and, anyway, a direct study of it offers some advantages with respect to the indications that the methods of Section 26 could yield.

The following hypothesis will be admitted.

There exists a domain $A_0 \supset \bar{A}$, such that a linear operator T_0 is defined in $\mathfrak{L}^2(A_0)$, which possesses the following properties:

(1) T_0 is a continuous operator mapping $\mathfrak{L}^2(A_0)$ into $H_\nu(A_0)$;

(2) The function $u = T_0 f$ solves the equation $Lu = f$;

(3) If T_0^* is the adjoint operator to T_0 in the space $\mathfrak{L}^2(A_0)$, i.e. if one has

$$(T_0 f, g)_{0,A_0} = (f, T_0^* g)_{0,A_0} \qquad f \in \mathfrak{L}^2(A_0), \qquad g \in \mathfrak{L}^2(A_0),$$

then, for $g \in \mathfrak{L}^2(A_0)$, one has $T_0^* g \in H_\nu(A_0)$ and $L^* T_0^* g = g$.

The hypothesis just assumed is obviously satisfied by the case of the classical biharmonic problem. For, it is enough to assume

$$T_0 f = \int_{A_0} s(x, y) f(y) \, dy$$

with

$$s(x, y) \begin{cases} = \dfrac{1}{2\pi} \log |x - y| & \text{if} \quad r = 2, \\[2ex] = -\dfrac{1}{\omega_r} |x - y|^{2-r} & \text{if} \quad r > 2 \end{cases}$$

[$\omega_r = (r-2)\tilde{\omega}_r$; $\tilde{\omega}_r$ is the hypersurface measure of the unit hypersphere of X^r].

Let P_A be the orthogonal projector of $\mathfrak{L}^2(A_0)$, mapping a function $f \in \mathfrak{L}^2(A_0)$ into the function vanishing on $A_0 - \bar{A}$ and coinciding with f in A.

Set

$$T = T_0 P_A, \qquad T^* = T_0^* P_A.$$

One proves [70] that the solution of problem (1.18.1) is unique and is given by $u = Gf$ with

$$G = TT^* - TPT^*, \tag{1.18.2}$$

P being the (orthogonal) projector of $\mathfrak{L}^2(A)$ onto the manifold Ω of the solutions of the equation $L^* u = 0$ belonging to $\mathfrak{L}^2(A)$.

Such a structure of the Green operator G is similar to the one obtained in Section 16; however it has the advantage of involving only the space $\mathfrak{L}^2(A)$. Let $\{\omega_k\}$ be a complete system in Ω and let Ω_ρ be the ρ-dimensional linear manifold spanned by $\omega_1, \ldots, \omega_\rho$.

Let P_ρ be the orthogonal projector of $\mathfrak{L}^2(A)$ onto Ω_ρ. Setting

$$G_\rho = TT^* - TP_\rho T^*,$$

one has $G_\rho > G_{\rho+1}$ and $\lim_{\rho\to\infty} \| G - G_\rho \| = 0$.

All considerations reported at the end of Section 16 may, thus, be applied again.

An interesting remark is that, assuming $l = 1, r = 2$,

$$L = \frac{\partial}{\partial x_1} - i\frac{\partial}{\partial x_2}$$

and hence

$$L^* = -\left(\frac{\partial}{\partial x_1} + i\frac{\partial}{\partial x_2}\right)$$

problem (1.18.1) reduces to the following one:

$$\Delta_2 u + f = 0 \text{ in A}, \qquad u = 0 \text{ on } \partial A.$$

That is, the classical harmonic problem is a particular case of the biharmonic problem, as defined above.

Notice that, in such a case, we may assume as T_0 the following operator:

$$T_0 f = \frac{1}{2\pi}\int_{A_0} \frac{f(\zeta)}{\bar z - \bar\zeta}\, d\xi_1\, d\xi_2 \qquad (z = x_1 + ix_2, \zeta = \xi_1 + i\xi_2).$$

Therefore one has:

$$Tf = \frac{1}{2\pi}\int_A \frac{f(\zeta)}{\bar z - \bar\zeta}\, d\xi_1\, d\xi_2, \qquad T^* f = \frac{1}{2\pi}\int_A \frac{f(\zeta)}{\zeta - z}\, d\xi_1\, d\xi_2. \tag{1.18.3}$$

In this particular case we get an illuminating interpretation of the structure formula (1.18.2) for the Green operator: the operators T and T^* are given by (1.18.3) and P is the projector of $\mathfrak{L}^2(A)$ onto the space of the functions holomorphic in A and belonging to $\mathfrak{L}^2(A)$.

The theory expounded in this Section has been recently extended by Romano [185] to more general situations.

19. The Structure Theorem of the Green Operator and the Orthogonal Invariants Method for the Clamped Plate Problem

As we did for the theory expounded in Section 16, we want to give a significant illustration also for the one sketched in the former Section, by means of a detailed investigation of a classical particular case: namely, the problem of the elastic plate clamped along its boundary, which is translated analytically into the biharmonic problem in two variables.

Let A be a bounded and properly regular domain belonging to X^2. The first step of the method consists in constructing the Green operator of the biharmonic problem

$$\begin{cases} \Delta_2 \Delta_2 u = f & \text{in } A, \\ u = \dfrac{\partial u}{\partial n} = 0 & \text{on } \partial A. \end{cases} \tag{1.19.1}$$

Given f, supposed real and belonging to $\mathfrak{L}^2(A)$, the solution will be sought within the class of the (real) functions belonging to $H_2(A) \cap H_4(B)$ for any domain B such that $\bar{B} \subset A$. Notice that in this Section the function spaces that occur will be supposed real.

Let T be a linear operator which possesses the following properties:

(1) T is defined in $\mathfrak{L}^2(A)$;

(2) Tv is defined in X^2 and belongs to $H_2(B)$, for any domain B of X^2, Tv being a continuous operator from $\mathfrak{L}^2(A)$ to $H_2(B)$;

(3) T, considered as an operator from $\mathfrak{L}^2(A)$ into $\mathfrak{L}^2(A)$, is symmetric;

(4) it turns out that $\Delta_2 Tw = w$.

The operator T may be constructed in several ways. Let us indicate two ways. The first one is the following:

$$Tw = \frac{1}{2\pi} \int_A w(y) \log |x - y| \, dy$$

$$[x \equiv (x_1, x_2), y \equiv (y_1, y_2); dy = dy_1 dy_2].$$

The second method yields Tw through a series expansion. In order to simplify the formalism, suppose, without loss of generality, A to be contained in the domain

$$0 < x_1 < \pi, \qquad 0 < x_2 < \pi.$$

One then has:

$$Tw = \frac{4}{\pi^4} \int_A w(y) K(x, y) \, dy,$$

where we have set

$$K(x_1, x_2; y_1, y_2) = -\sum_{l,m}^{1,\infty} \frac{\sin lx_1 \sin mx_2 \sin ly_1 \sin my_2}{l^2 + m^2}.$$

In what follows we will denote by $s(x, y)$ any one of the two functions

$\frac{1}{2\pi} \log |x-y|, \frac{4}{\pi^2} K(x,y)$.

Suppose we wish to determine w in such a way that the function $u = Tw$ is the solution (easily proved to be unique) of problem (1.19.1) that we are seeking. The function $u = Tw$ satisfies the boundary conditions required by (1.19.1), if $Pw = 0$, P being the orthogonal projector of $\mathfrak{L}^2(A)$ onto its subspace $\Omega(A)$, formed of the square integrable harmonic functions in A. After this remark, it is easily seen that it suffices to assume $w = Tf - PTf$. Hence u is given by $u = Gf$, with

$$G = T^2 - TPT.$$

The Green operator of problem (1.19.1) has, thus, been constructed. It is easily proved [70] that $G \in \mathfrak{C}^1$. Introducing a complete system $\{\omega_i\}$ of linearly independent harmonic functions, complete in $\Omega(A)$, let $\Omega_\rho(A)$ be the manifold spanned by $\omega_1, \ldots, \omega_\rho$ and P_ρ be the orthogonal projector of $\mathfrak{L}^2(A)$ onto $\Omega_\rho(A)$. Setting

$$G_\rho = T^2 - TP_\rho T,$$

one has $G_\rho > G$ and $\lim_{\rho\to\infty} \| G - G_\rho \| = 0$. On the other hand, G_ρ turns out to be an integral operator whose kernel $g_\rho(x,y)$ is explicitly computable. Precisely, defining $\{\alpha_{ih}^{(\rho)}\}$ $(i, h = 1, \ldots, \rho)$ as the inverse of the matrix $\left\{\int_A \omega_i \omega_h \, dx\right\}$ $(i, h = 1, \ldots, \rho)$,

one has:

$$g_\rho(x,y) = \int_A s(x,\xi)s(\xi,y)d\xi - \sum_{s,h}^{1,\rho} \alpha_{ih}^{(\rho)} \int_A s(x,\xi)\omega_i(\xi)d\xi \int_A s(y,\xi)\omega_h(\xi)\, d\xi.$$

Therefore by means of (1.10.2) the orthogonal invariants $\mathfrak{J}_s^n(G_\rho)$ may be computed. For instance, in the simplest case $s = n = 1$,

$$\mathfrak{J}_1^1(G_\rho) = \int_A \int_A |s(x,y)|^2 \, dx\, dy$$

$$-\sum_{i,h}^{1,\rho} \alpha_{ih}^{(\rho)} \int_A \left[\int_A s(x,y)\,\omega_i(y)\,dy \int_A s(x,y)\,\omega_h(y)\,dy\right] dx.$$

This allows to the actual computation of the terms of a sequence yielding lower approximations to the eigenvalue λ_h relative to the clamped plate fixed along the boundary.

To give an example, assume $s = n = 1$, $\rho = \nu$.

One then has

$$\tau_k^{(\nu)} \leq \lambda_k \leq \lambda_k^{(\nu)};$$

$\lambda_1^{(1)} \leqslant \ldots \leqslant \lambda_\nu^{(\nu)}$ are the roots of the secular equation:

$$\det\left\{\int_A \Delta_2 w_i \, \Delta_2 w_j \, dx - \lambda \int_A w_i w_j\right\} = 0,$$

$w_1, \ldots, w_\nu$ being linearly independent functions vanishing on ∂A together with their first derivatives; we have furthermore set:

$$\tau_k^{(\nu)} = \left\{\int_A \int_A |s(x,y)|^2 \, dx \, dy\right.$$

$$\left. - \sum_{i,h}^{1,\nu} \alpha_{ih}^{(\nu)} \int_A \left[\int_A s(x,y)\, \omega_i(y)\, dy \int_A s(x,y)\, \omega_h(y)\, dy\right] dx - \sum_i^{1,\nu}{}^{(k)} \frac{1}{\lambda_i^{(\nu)}}\right\}^{-1}.$$

Using higher index orthogonal invariants less simple and elegant formulae are obtained, which, however, in general, yield better numerical results.

20. Computation of the Eigenvalues of the Square Clamped Plate

The method illustrated in the former Section has been applied to the case in which A is square: $|x| < \frac{\pi}{2}$, $|y| < \frac{\pi}{2}$. This is a classical and difficult eigenvalue problem: Ritz, Weinstein and other authors have all experimented with it to examine the effectiveness of their methods.

Weinstein [27] obtained in 1936 the following bounds on the first four eigenvalues:

$$13.294 < \lambda_1 < 13.37$$

$$50.41 < \lambda_2 = \lambda_3 < 55.76$$

$$112.36 < \lambda_4 < 134.56.$$

Such numerical results, taking into account the considerable difficulty of the problem as well as the fact that the electronic computers did not then exist, are to be considered of great value. Therefore it must be considered as irrelevant that the lower bound obtained for λ_1 is slightly larger than the actual value of λ_1.

In 1950 Aronszajn [29], applying again the Weinstein method and using an electronic computer (employed for the first time with a complicated eigenvalue problem!), obtained the following bounds:

	Lower bounds	Upper bounds
λ_1	13.282	13.3842
λ_2	55.24	56.561
λ_3	55.24	56.561
λ_4	120.007	120.074
λ_5	177.67	182.14
λ_6	178.3	184.5
λ_7	277.42	301.55
λ_8	277.42	301.55
λ_9	454	477
λ_{10}	454	477
λ_{11}	488	548
λ_{12}	600.840	621.852
λ_{13}	601.569	646.939

Bazley D.W. Fox and Stadter [107], always by means of the intermediate problems method, have subsequently sharpened the Aronszajn results, obtaining, for instance, for the first eigenvalue

$$13.2916 < \lambda_1 < 13.29387$$

and altogether yielding bounds for 15 eigenvalues of the problem.

The problem of the square clamped plate has been deeply investigated [108], where the methods described in the former Sections and, in particular, in Section 19, have been applied.

Denote by $H^{(a_1 a_2)}$ ($a_i = 0, 1$) the subspace of $\mathfrak{L}^2$(A) formed of all functions symmetric with respect to the x_i axis, if $a_i = 0$ and antisymmetric with respect to the same axis if $a_i = 1$.

Let $H^{(a_1, a_2, a_3)}$ be the subspace of $H^{(a_1, a_2)}$ formed by the symmetric (antisymmetric) functions with respect to the line $x_1 - x_2 = 0$ if $a_3 = 0$ (if $a_3 = 1$).

The space $\mathfrak{L}^2$(A) can be decomposed as a direct sum of six subspaces; each of them is an invariant subspace for the problem (1.19.1) (*see* Section 15).

$$\mathfrak{L}^2(A) = H^{(000)}(A) \oplus H^{(001)}(A) \oplus H^{(110)}(A) \oplus H^{(111)}(A) \oplus \oplus H^{(01)}(A) \oplus H^{(10)}(A). \tag{1.20.1}$$ [19]

Let H be any one of the six subspaces operating the decomposition (1.20.1) and let Q be the orthogonal projector of $\mathfrak{L}^2$(A) onto H. Define P as in the preceding Section Set $P^{(H)} = QPQ$ and let G be the Green operator of problem (1.19.1). Let

(19) It is interesting to remark that Aronszajn, in his study of the same problem, did not notice the possibility of decomposing $\mathfrak{L}^2$(A) into six invariant subspaces, but only obtained the decomposition of $\mathfrak{L}^2$(A) into the four subspaces $H^{(00)}$, $H^{(01)}$, $H^{(10)}$, $H^{(11)}$.

$$G^{(H)} = (QTQ)^2 - QTP^{(H)}TQ,$$

T being the operator introduced in Section 19, whose kernel is $\frac{4}{\pi^2}$ K (x, y). The problem is equivalent to the system formed by the six following eigenvalue problems:

$$u - \lambda G^{(H)} u = 0, \qquad u \in H \tag{1.20.2}$$

for all this the H's in the right-hand side of (1.20.1).

Let $\{w_k^{(H)}\}$ be a system of linearly independent functions, belonging to $H \cap \mathfrak{U}(A)$ and satisfying the boundary conditions of problem (1.20.1).

Let $\lambda_1^{(n)} \leqslant \ldots \leqslant \lambda_n^{(n)}$ be the solutions of the secular equation

$$\det\left\{\int_A \Delta_2 w_i^{(H)} \Delta_2 w_j^{(H)} dx - \lambda \int_A w_i^{(H)} w_j^{(H)} dx\right\} = 0 \qquad (i, j = 1, \ldots, n).$$

For the k-th eigenvalue of (1.20.2) one has:

$$\tau_k^{(n,\rho)} \leqslant \lambda_k^{(H)} \leqslant \lambda_k^{(n)}$$

with

$$\tau_k^{(n,\rho)} = \left\{\mathfrak{I}_1^2(G_\rho^{(H)}) - \sum_{h=1}^{n}{}^{(k)} [\lambda_h^{(n)}]^{-2}\right\}^{-1/2},$$

and

$$G_\rho^{(H)} = (QTQ)^2 - QTP_\rho^{(H)}TQ$$

the meaning of P_ρ being already specified in the former Sections.

It must be kept in mind that, for any k, it turns out that

$$\lambda_k^{(H^{(01)})} = \lambda_k^{(H^{(10)})}$$

as it is easily verified.

The method yields the following bounds on the eigenvalues relative to any invariant subspace:

$H^{(000)}$

	Lower bounds	Upper bounds
$\lambda_1^{(1)}$	13.29376	13.29378
$\lambda_2^{(1)}$	179.408	179.431
$\lambda_3^{(1)}$	496.55	497.03
$\lambda_4^{(1)}$	977.64	981.25
$\lambda_5^{(1)}$	1577	1593
$\lambda_6^{(1)}$	3120	3244
$\lambda_7^{(1)}$	3155	3284
$\lambda_8^{(1)}$	4037	4317
$\lambda_9^{(1)}$	5853	6817
$\lambda_{10}^{(1)}$	6701	8276

$H^{(001)}$

	Lower bounds	Upper bounds
$\lambda_1^{(2)}$	177.7193	177.7401
$\lambda_2^{(2)}$	976.13	979.59
$\lambda_3^{(2)}$	1569	1584
$\lambda_4^{(2)}$	3158	3282
$\lambda_5^{(2)}$	4038	4306
$\lambda_6^{(2)}$	5865	6791
$\lambda_7^{(2)}$	6774	8330
$\lambda_8^{(2)}$	7555	9931

$H^{(110)}$

	Lower bounds	Upper bounds
$\lambda_1^{(3)}$	120.2143	120.2232
$\lambda_2^{(3)}$	605.792	606.920
$\lambda_3^{(3)}$	1401.5	1415.7
$\lambda_4^{(3)}$	2111.8	2161.2
$\lambda_5^{(3)}$	3306	3506
$\lambda_6^{(3)}$	5037	5842
$\lambda_7^{(3)}$	5412	6451
$\lambda_8^{(3)}$	6200	7931

$H^{(111)}$

	Lower bounds	Upper bounds
$\lambda_1^{(4)}$	601.488	601.983
$\lambda_2^{(4)}$	2133.1	2155.6
$\lambda_3^{(4)}$	3398	3491
$\lambda_4^{(4)}$	5429	5834
$\lambda_5^{(4)}$	6970	7894
$\lambda_6^{(4)}$	9366	12071

$H^{(01)}$ and $H^{(10)}$

	Lower bounds	Upper bounds	
$\lambda_1^{(i)}$	55.2982	55.2994	
$\lambda_2^{(i)}$	279.35	279.50	
$\lambda_3^{(i)}$	454.37	454.99	
$\lambda_4^{(i)}$	896.8	901.6	
$\lambda_5^{(i)}$	1180	1191	$(i = 5,6)$
$\lambda_6^{(i)}$	1833	1875	
$\lambda_7^{(i)}$	2171	2242	
$\lambda_8^{(i)}$	2560	2677	
$\lambda_9^{(i)}$	3371	3652	
$\lambda_{10}^{(i)}$	4154	4716	
$\lambda_{11}^{(i)}$	4556	5329	
$\lambda_{12}^{(i)}$	4582	5372	

Using Theorem 1.XI the following table, yielding bounds on the first 45 eigenvalues of (1.19.1), is obtained.

Given the high number of eigenvalues (as many as 45!) for which it has been possible to get bounds, it has been thought to compare such bounds with the values obtained through the asymptotic formula of Courant and Pleijel, recalled in Section 6 (*see* (1.6.13)) and extended by Agmon (*see* Section 7).

Such a comparison, *which was impossible to make before*, exhibited surprisingly catastrophic results for the convergence of the asymptotic formula. For, it may be seen from the following table that, while on λ_{45} the bounds give

$$4582 < \lambda_{45} < 5372,$$

the asymptotic formula (1.6.13) yields

$$\lambda_{45} \simeq 3282.8.$$ [20]

---

[20] After the publication of the following table, Clark [109], *taking into account the fact that* A *is a square*, improved the asymptotic formula (1.6.13) of Courant and Pleijel and obtained

$$\lambda_{45} \simeq 4225.$$

Notice that, even after Clark's improvements, the value given by the asymptotic formula lies outside the interval obtained above.

	Lower bounds	Upper bounds	Asymptotic values
λ_1	13.29376	13.29378	1.6211
λ_2	55.2982	55.29934	6.4845
λ_3	55.2982	55.29934	14.590
λ_4	120.2143	120.2232	25.938
λ_5	177.7113	177.7401	40.528
λ_6	179.408	179.431	58.361
λ_7	279.35	279.50	79.435
λ_8	279.35	279.50	103.75
λ_9	454.37	454.99	131.31
λ_{10}	454.37	454.99	162.11
λ_{11}	496.55	497.03	196.15
λ_{12}	601.488	601.983	233.44
λ_{13}	605.792	606.920	273.97
λ_{14}	896.8	901.6	317.74
λ_{15}	896.8	901.6	364.75
λ_{16}	976.13	979.59	415.01
λ_{17}	977.64	981.25	468.50
λ_{18}	1180	1191	525.24
λ_{19}	1180	1191	585.23
λ_{20}	1401	1415	648.45
λ_{21}	1569	1584	714.92
λ_{22}	1577	1593	784.63
λ_{23}	1833	1875	857.58
λ_{24}	1833	1875	933.77
λ_{25}	2111.8	2155.6	1013.2
λ_{26}	2133.1	2161.2	1095.8
λ_{27}	2171	2242	1181.8
λ_{28}	2171	2242	1270.9
λ_{29}	2560	2677	1363.3
λ_{30}	2560	2677	1459.0
λ_{31}	3120	3244	1557.9
λ_{32}	3155	3282	1660.0
λ_{33}	3158	3284	1765.4
λ_{34}	3306	3491	1874.0
λ_{35}	3371	3506	1985.8
λ_{36}	3371	3652	2100.9
λ_{37}	3398	3652	2219.3
λ_{38}	4037	4306	2340.9
λ_{39}	4038	4317	2465.7

	Lower bounds	Upper bounds	Asymptotic values
λ_{40}	4154	4716	2593.8
λ_{41}	4154	4716	2725.1
λ_{42}	4556	5329	2859.6
λ_{43}	4556	5329	2997.4
λ_{44}	4582	5372	3138.5
λ_{45}	4582	5372	3282.8

21. The Eigenvalues of the Circular Clamped Plate

The discovery of the inadequacy of the asymptotic formulae in yielding acceptable approximate values, also for higher index eigenvalues, led the writer and his coworkers to consider a case where it is possible to compute both upper and lower bounds to a large number of eigenvalues with a good approximation, so that a comparison could be made of these values with those given by the asymptotic formulae. Therefore the case has been considered [110] of a circular plate clamped along the boundary, i.e. the eigenvalue problem (1.19.1), where A is a circular domain of radius 1. In this case the decomposition of $\mathfrak{L}^2(A)$ into invariant subspaces is made through a denumerable infinity of invariant subspaces, labelled in the following way:

$\mathfrak{L}^2_0, \mathfrak{L}^2_{1k}, \mathfrak{L}^2_{2k}$ $(k = 1, 2, \ldots)$.

Introducing a system of polar coordinates ρ, θ, the pole being placed in the centre of A, $\mathfrak{L}^2_0$ is the space of those functions of $\mathfrak{L}^2(A)$ depending only on ρ, whereas $\mathfrak{L}^2_{1k}$ and $\mathfrak{L}^2_{2k}$ are the spaces formed, respectively, by functions of the following type:

$$\varphi(\rho) \cos k\theta, \qquad \varphi(\rho) \sin k\theta.$$

Any one of these spaces is an invariant subspace for the problem (1.19.1), in the particular case we are now considering.

Denoting by $v(\rho)$ a function of ρ, set:

$$L_k v = \frac{d^2 v}{d\rho^2} + \frac{1}{\rho}\frac{dv}{d\rho} - \frac{k^2}{\rho^2} v.$$

Problem (1.19.1) is equivalent to the following sequence of eigenvalue problems:

$$L_k L_k v - \lambda v = 0, \qquad v(1) = v'(1) = 0 \tag{1.21.1}$$

$$v, v', v'' \qquad \text{continuous at } \rho = 0$$

$$(k = 0, 1, 2, \ldots).$$

The eigenvalues corresponding to $k > 0$ are double eigenvalues for problem (1.19.1).

The upper approximations are, as usual, obtained through the Rayleigh–Ritz method by means of the following functions:

$$w_i(\rho) = (1 - \rho^2)\,\rho^{k+2i}.$$

The orthogonal invariants of the Green operator G_k of problem (1.21.1) are elementary to compute, and in particular, one has:

$$\mathfrak{I}_1^2(G_k) = \frac{1}{2^8}\,\frac{(1-\delta_{ok})(5k+17)}{(k+1)^2(k+2)^2(k+3)^2\,(k+4)\,(k+5)}\,.$$

$(\delta_{ok} = 1$ for $k = 0$; $\delta_{ok} = 0$ for $k \neq 0)$.

The following tables show the bounds on the first eigenvalues for $k = 0, 1, \ldots, 20$.

	$k = 0$		$k = 1$	
	Lower bounds	Upper bounds	Lower bounds	Upper bounds
λ_1	104.3631051	104.3631056	452.00448	452.00452
λ_2	1581.742	1581.745	3700.11	3700.13
λ_3	7939.38	7939.55	14418.2	14419.1
λ_4	25017.2	25022.3	39606.2	39622.3
λ_5	60939.5	61012.2	88482.2	88661.1
λ_6	125786	126430	171901	173225
λ_7	230123	234133	300129	307340
λ_8	380355	399323	476778	507392
λ_9	569823	640349	689901	794004

	$k = 2$		$k = 3$	
	Lower bounds	Upper bounds	Lower bounds	Upper bounds
λ_1	1216.4072	1216.4076	2604.061	2604.065
λ_2	7154.14	7154.23	12325.4	12325.8
λ_3	23656.3	23659.1	36207.4	36215.6
λ_4	58870.7	58913.3	83526.1	83625.1
λ_5	123047	123437	165470	166244
λ_6	227594	230089	293711	298098
λ_7	381914	394063	476150	495553
λ_8	585987	632954	708346	777466

	$k = 4$		$k = 5$	
	Lower bounds	Upper bounds	Lower bounds	Upper bounds
λ_1	4853.31	4853.33	8233.49	8233.57
λ_2	19629.1	19630.3	29513.3	29516.3
λ_3	52658.5	52678.8	73627.7	73673.3
λ_4	114314	114523	152001	152404
λ_5	216597	218019	277274	279738
λ_6	371076	378366	460483	472040
λ_7	583460	613097	704421	748019

	$k = 6$		$k = 7$	
	Lower bounds	Upper bounds	Lower bounds	Upper bounds
λ_1	13044.2	13044.5	19615.1	19615.8
λ_2	42457.8	42465.1	58973.7	58989.9
λ_3	99763.2	99857.0	131741	131922
λ_4	197374	198104	251235	252489
λ_5	348349	352407	430663	437070
λ_6	562700	580302	678459	704371

	$k = 8$		$k = 9$	
	Lower bounds	Upper bounds	Lower bounds	Upper bounds
λ_1	28304.7	28306.3	39500.6	39504.1
λ_2	79602.4	79635.6	104914	104979
λ_3	170265	170590	216062	216620
λ_4	314402	316460	387704	390951
λ_5	525050	534802	632331	646714
λ_6	808467	845496		

	$k = 10$		$k = 11$	
	Lower bounds	Upper bounds	Lower bounds	Upper bounds
λ_1	53618.9	53626.1	71103.5	71117.8
λ_2	135509	135626	172014	172215
λ_3	269882	270798	332495	333947
λ_4	471976	976928	568059	575391
λ_5	753313	773948	888788	917682

	$k = 12$		$k = 13$	
	Lower bounds	Upper bounds	Lower bounds	Upper bounds
λ_1	92426.2	92452.6	118085	118133
λ_2	215080	215414	265387	265921
λ_3	404689	406917	487270	490593
λ_4	676795	687371	799027	813933

	$k = 14$		$k = 15$	
	Lower bounds	Upper bounds	Lower bounds	Upper bounds
λ_1	148607	148687	184542	184674
λ_2	323636	324465	390553	391804
λ_3	581056	585887	686877	693747
λ_4	935594	956172		

	$k = 16$		$k = 17$	
	Lower bounds	Upper bounds	Lower bounds	Upper bounds
λ_1	226468	226678	274986	275311
λ_2	466883	468724	553390	556044
λ_3	805574	815148	937997	951097

	$k = 18$		$k = 19$		$k = 20$	
	Lower bounds	Upper bounds	Lower bounds	Upper bounds	Lower bounds	Upper bounds
λ_1	330725	331214	394333	395054	466485	467526
λ_2	650861	654609	760097	765295	881916	889004

Several textbooks contain the following tables, due to Carrington [111], where $\nu_j = \lambda_j^{1/4}$. In these tables the ν_j's are obtained as zeros of the following transcendental function, expressed by means of Bessel functions:

$$iJ_k(\nu)\, J_{k+1}(i\nu) - J_k(i\nu)\, J_{k+1}(\nu).$$

	$k = 0$	$k = 1$	$k = 2$	$k = 3$
ν_1	3.1961	4.6110	5.9056	7.1433
ν_2	6.3064	7.7993	9.1967	10.537
ν_3	9.4395	10.958	12.402	13.795
ν_4	12.577	14.108	15.579	
ν_5	15.716			

The tables reported above yield:

	$k = 0$		$k = 1$	
	Lower bounds	Upper bounds	Lower bounds	Upper bounds
ν_1	3.19622	3.19623	4.61089	4.61090
ν_2	6.30643	6.30644	7.79926	7.79928
ν_3	9.43945	9.43950	10.9579	10.9581
ν_4	12.5764	12.5772	14.1072	14.1087
ν_5	15.7117	15.7165	17.2470	17.2558

	$k = 2$		$k = 3$	
	Lower bounds	Upper bounds	Lower bounds	Upper bounds
ν_1	5.90567	5.90568	7.14352	7.14354
ν_2	9.19685	9.19689	10.5366	10.5367
ν_3	12.4018	12.4023	13.7942	13.7951
ν_4	15.5766	15.5795	17.0002	17.0053

Comparing with the above tables, we see that some values of Carrington's lie outside the intervals obtained by us.

Using Theorem 1.XI, the following table is constructed, which yields bounds on the first 160 eigenvalues of the circular clamped plate. They are compared with the asymptotic values given by (1.6.13).

	Lower bounds	Upper bounds	Asymptotic values
λ_1	104.363	104.364	16
λ_2	452.004	452.005	64
λ_3	452.004	452.005	144
λ_4	1216.40	1216.41	256
λ_5	1216.40	1216.41	400
λ_6	1581.74	1581.75	576
λ_7	2604.06	2604.07	784
λ_8	2604.06	2604.07	1024
λ_9	3700.11	3700.13	1296
λ_{10}	3700.11	3700.13	1600
λ_{11}	4853.31	4853.33	1936
λ_{12}	4853.31	4953.33	2304
λ_{13}	7154.14	7154.23	2704
λ_{14}	7154.14	7154.23	3136
λ_{15}	7939.38	7939.55	3600

	Lower bounds	Upper bounds	Asymptotic values
λ_{16}	8233.49	8233.57	4096
λ_{17}	8233.49	8233.57	4624
λ_{18}	12325.4	12325.76	5184
λ_{19}	12325.4	12325.76	5776
λ_{20}	13044.2	13044.5	6400
λ_{21}	13044.2	13044.5	7056
λ_{22}	14418.2	14420.0	7744
λ_{23}	14418.2	14420.0	8464
λ_{24}	19615.1	19615.8	9216
λ_{25}	19615.1	19615.8	10000
λ_{26}	19629.1	19630.3	10816
λ_{27}	19629.1	19630.3	11664
λ_{28}	23656.3	23659.1	12544
λ_{29}	23656.3	23659.1	13456
λ_{30}	25017.2	25022.3	14400
λ_{31}	28304.7	28306.3	15376
λ_{32}	28304.7	28306.3	16384
λ_{33}	29513.3	29516.3	17424
λ_{34}	29513.3	29516.3	18496
λ_{35}	36207.4	36215.6	19600
λ_{36}	36207.4	36215.6	20736
λ_{37}	39500.6	39504.1	21904
λ_{38}	39500.6	39504.1	23104
λ_{39}	39606.2	39622.3	24336
λ_{40}	39606.2	39622.3	25600
λ_{41}	42457.8	42465.1	26896
λ_{42}	42457.8	42465.1	28224
λ_{43}	52658.5	52678.8	29584
λ_{44}	52658.5	52678.8	30976
λ_{45}	53618.9	53626.1	32400
λ_{46}	53618.9	53626.1	33856
λ_{47}	58870.7	58913.3	35344
λ_{48}	58870.7	58913.3	36804
λ_{49}	58973.7	58989.9	38416
λ_{50}	58973.7	58989.9	40000
λ_{51}	60939.5	61012.2	41616
λ_{52}	71103.5	71117.8	43264
λ_{53}	71103.5	71117.8	44944
λ_{54}	73627.7	73673.3	46656
λ_{55}	73627.7	73673.3	48400
λ_{56}	79602.4	79635.6	50176

	Lower bounds	Upper bounds	Asymptotic values
λ_{57}	79602.4	79635.6	51984
λ_{58}	83526.1	83625.1	53824
λ_{59}	83526.1	83625.1	55696
λ_{60}	88482.2	88661.1	57600
λ_{61}	88482.2	88661.1	59536
λ_{62}	92426.2	92452.6	61504
λ_{63}	92426.2	92452.6	63504
λ_{64}	99763.2	99857.0	65536
λ_{65}	99763.2	99857.0	67600
λ_{66}	104914	104979	69696
λ_{67}	104914	104979	71824
λ_{68}	114314	114523	73984
λ_{69}	114314	114523	76176
λ_{70}	118085	118133	78400
λ_{71}	118085	118133	80656
λ_{72}	123047	123437	82944
λ_{73}	123047	123437	85264
λ_{74}	125786	126430	87616
λ_{75}	131741	131922	90000
λ_{76}	131741	131922	92416
λ_{77}	135509	135626	94864
λ_{78}	135509	135626	97344
λ_{79}	148607	148687	99856
λ_{80}	148607	148687	102400
λ_{81}	152001	152404	104976
λ_{82}	152001	152404	107584
λ_{83}	165470	166244	110224
λ_{84}	165470	166244	112896
λ_{85}	170265	170590	115600
λ_{86}	170265	170590	118336
λ_{87}	171901	172215	121104
λ_{88}	171901	172215	123904
λ_{89}	172014	173225	126736
λ_{90}	172014	173225	129600
λ_{91}	184542	184674	132496
λ_{92}	184542	184674	135424
λ_{93}	197374	198104	138384
λ_{94}	197374	198104	141376
λ_{95}	215080	215414	144400
λ_{96}	215080	215414	147456
λ_{97}	216062	216620	150544

	Lower bounds	Upper bounds	Asymptotic values
λ_{98}	216062	216620	153664
λ_{99}	216597	218019	156816
λ_{100}	216597	218019	160000
λ_{101}	226468	226678	163216
λ_{102}	226468	226678	166464
λ_{103}	227594	230089	169744
λ_{104}	227594	230089	173056
λ_{105}	230123	234133	176400
λ_{106}	251235	252489	179776
λ_{107}	251235	252489	183184
λ_{108}	265387	265921	186624
λ_{109}	265387	265921	190096
λ_{110}	269882	270798	193600
λ_{111}	269882	270798	197136
λ_{112}	274986	275311	200704
λ_{113}	274986	275311	204304
λ_{114}	277274	279738	207936
λ_{115}	277274	279738	211600
λ_{116}	293711	298098	215296
λ_{117}	293711	298098	219024
λ_{118}	300129	307340	222784
λ_{119}	300129	307340	226576
λ_{120}	314402	316460	230400
λ_{121}	314402	316460	234256
λ_{122}	323636	324465	238144
λ_{123}	323636	324465	242064
λ_{124}	330725	331214	246016
λ_{125}	330725	331214	250000
λ_{126}	332495	333947	254016
λ_{127}	332495	333947	258064
λ_{128}	348349	352407	262144
λ_{129}	348349	352407	266256
λ_{130}	371076	378366	270400
λ_{131}	371076	378366	274576
λ_{132}	380355	390951	278784
λ_{133}	381914	390951	283024
λ_{134}	381914	391804	285296
λ_{135}	387704	391804	291600
λ_{136}	387704	394063	295936
λ_{137}	390553	394063	300304
λ_{138}	390553	395054	304704

	Lower bounds	Upper bounds	Asymptotic values
λ_{139}	394333	395054	309136
λ_{140}	394333	399323	313600
λ_{141}	404689	406917	318096
λ_{142}	404689	406917	322624
λ_{143}	430663	437070	327184
λ_{144}	430663	437070	331776
λ_{145}	460483	467526	336400
λ_{146}	460483	467526	341056
λ_{147}	466485	468724	345744
λ_{148}	466485	468724	350464
λ_{149}	466882	472040	355216
λ_{150}	466882	472040	360000
λ_{151}	471976	476928	364816
λ_{152}	471976	476928	369664
λ_{153}	476150	490593	374544
λ_{154}	476150	490593	379456
λ_{155}	476778	495553	384400
λ_{156}	476778	495553	389376
λ_{157}	487270	507392	394384
λ_{158}	487270	507392	399424
λ_{159}	525050	534801	404496
λ_{160}	525050	534801	409600

In this case too (although as many as 160 eigenvalues have been computed!) the asymptotic formula of Courant–Pleijel–Agmon shows its inadequacy.(21)

The graph of Fig. 1.5 compares (against the increasing of the eigenvalue index, reported in abscissa) the behaviour of the upper bounds ϵ_k, normalized to the constant value 1, the behaviour of the lower bounds, consequently expressed by δ_k/ϵ_k and the behaviour of the asymptotic value expressed by α_k/ϵ_k, α_k being given (*see* (1.6.13)) by

$$\alpha_k = \frac{16k^2}{\pi^2}.$$

(21) Also for this case Clark [109], *taking into account the fact that A is a circular domain,* suggested an asymptotic formula improving (1.6.13). Therefore he obtains

$$\lambda_{160} \simeq 478000.$$

Even this value, however, cannot be considered satisfactory, since it turns out that

$$525050 < \lambda_{160} < 534801.$$

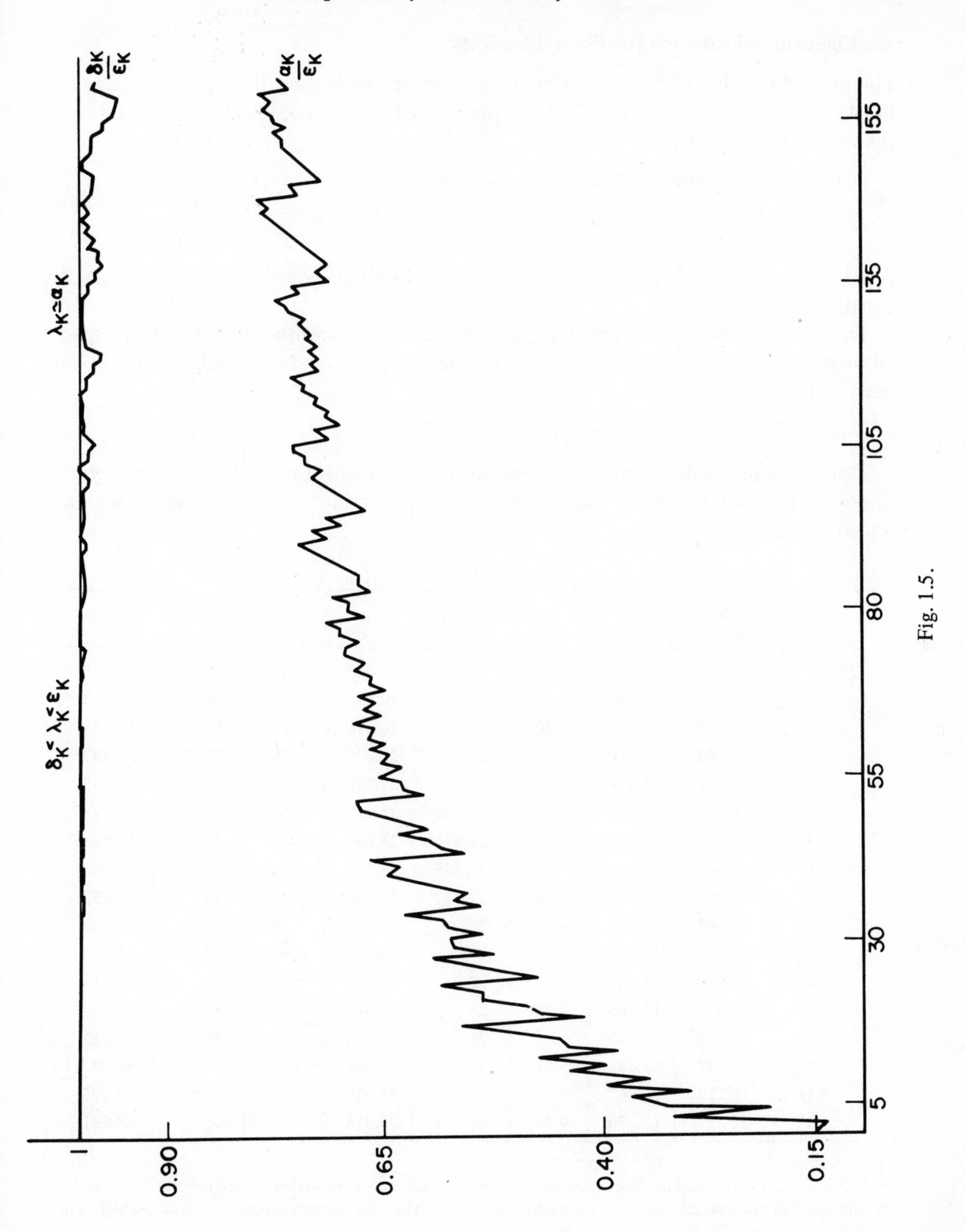

Fig. 1.5.

22. Eigenvalue Problems for Plane Elasticity

The procedures described in Section 17 have been applied by Bassotti in [112], [113], [114] to some eigenvalue problems of plane elasticity. Precisely, in [112] and in [113] the problem is considered

$$\begin{aligned} \Delta_2 w + \sigma \text{ grad div } w + \lambda w &= 0 \quad \text{in} \quad A, \\ w &= 0 \quad \text{on } \partial A. \end{aligned} \tag{1.22.1}$$

A is the square $|x| < \frac{\pi}{2}$, $|y| < \frac{\pi}{2}$ and w a real two-component vector; σ is a positive constant.

The space of the two-component vectors has been decomposed into six subspaces, all of them invariant subspaces for the problem (1.22.1).[22] Numerical results have been obtained for σ:

$$\sigma = 0.5, \qquad \sigma = 1, \qquad \sigma = 1.5, \qquad \sigma = 2, \qquad \sigma = 3.$$

The following tables contain the bounds to the eigenvalues obtained, through Theorem 1.XI, by means of those relative to the invariant subspaces, which we do not report here.

$\sigma = 0.5$

λ_k	Lower bounds	Upper bounds	λ_k	Lower bounds	Upper bounds	λ_k	Lower bounds	Upper bounds
$k = 1$	2.474	2.480	$k = 18$	17.997	19.033	$k = 35$	29	35.315
= 2	2.474	2.480	= 19	18	20.176	= 36	29	35.803
= 3	5.074	5.097	= 20	18	20.176	= 37	30.530	36.861
= 4	5.783	5.816	= 21	20	23.230	= 38	30.720	35.937
= 5	6.511	6.576	= 22	20	23.230	= 39	32	37.893
= 6	7.212	7.301	= 23	21.026	24.718	= 40	32	37.893
= 7	9.098	9.401	= 24	21.026	24.889	= 41	34	40.068
= 8	9.098	9.401	= 25	21.855	25.834	= 42	34	40.068
= 9	10.365	10.819	= 26	21.952	25.834	= 43	34	41.050
= 10	10.365	10.819	= 27	25	26.574	= 44	34	41.739
= 11	13.078	13.621	= 28	25	28.309	= 45	37	41.914
= 12	13.429	14.463	= 29	25.232	28.642	= 46	37	41.914
= 13	13.429	14.463	= 30	26	28.642	= 47	37	43.898
= 14	15.037	15.648	= 31	26	30.015	= 48	37	45.211
= 15	15.039	15.888	= 32	26	30.961	= 49	40	48.027
= 16	16.129	16.862	= 33	26	32.319	= 50	40	48.027
= 17	17	17.770	= 34	27.018	35.315	= 51	40	48.091

[22] The analysis performed by Bassotti in order to obtain such a decomposition is very interesting. In the next Section we shall describe the type of results achieved when considering the three-dimensional case which is of much greater complexity than the plane one.

λ_k	Lower bounds	Upper bounds	λ_k	Lower bounds	Upper bounds	λ_k	Lower bounds	Upper bounds
k = 52	40	50.079	k = 69	52	64.838	k = 86	65	79.574
= 53	41	50.079	= 70	52	65.140	= 87	65	80.953
= 54	41	50.639	= 71	53	65.703	= 88	65	81.518
= 55	41	52.098	= 72	53	65.791	= 89	65	81.518
= 56	41	52.498	= 73	53	65.791	= 90	65	82.528
= 57	45	52.607	= 74	53	66.583	= 91	68	82.528
= 58	45	55.552	= 75	58	68.616	= 92	68	85.236
= 59	45	55.552	= 76	58	68.616	= 93	68	85.238
= 60	45	55.870	= 77	58	71.137	= 94	68	85.417
= 61	50	56.261	= 78	58	72.870	= 95	72	87.891
= 62	50	56.956	= 79	61	72.927	= 96	72	88.669
= 63	50	56.956	= 80	61	72.927	= 97	73	88.669
= 64	50	57.959	= 81	61	72.980	= 98	73	91.220
= 65	50	61.693	= 82	61	74.644	= 99	73	91.446
= 66	50	61.693	= 83	65	75.192	= 100	73	91.725
= 67	52	62.108	= 84	65	75.192	= 101	74	91.725
= 68	52	63.788	= 85	65	76.848			

$\sigma = 1$			$\sigma = 1.5$			$\sigma = 2$		
λ_k	Lower bounds	Upper bounds	λ_k	Lower bounds	Upper bounds	λ_k	Lower bounds	Upper bounds
k = 1	2.926	2.932	k = 1	2.926	3.363	k = 1	3.244	3.779
= 2	2.926	2.932	= 2	2.926	3.363	= 2	3.244	3.779
= 3	5.113	5.147	= 3	5.136	5.179	= 3	5.136	5.202
= 4	6.476	6.575	= 4	7.075	7.296	= 4	7.585	7.983
= 5	7.792	8.001	= 5	8.797	9.223	= 5	9.515	10.113
= 6	9.142	9.492	= 6	9.623	9.989	= 6	9.623	10.113
= 7	9.623	9.817	= 7	9.623	9.989	= 7	9.623	10.189
= 8	9.623	9.817	= 8	10.763	11.589	= 8	11.931	13.600
= 9	11.931	12.305	= 9	11.931	13.767	= 9	11.931	15.018
= 10	11.931	12.305	= 10	11.931	13.767	= 10	12.083	15.018
= 11	13.091	14.160	= 11	13.210	14.800	= 11	13.477	15.629
= 12	15.453	16.463	= 12	15.410	16.653	= 12	14.970	16.742
= 13	16.129	17.258	= 13	16.129	17.478	= 13	15.410	17.688
= 14	16.206	18.436	= 14	16.838	20.669	= 14	17.121	21.776
= 15	17.256	18.448	= 15	17.256	20.694	= 15	17.256	21.776
= 16	17.256	18.448	= 16	17.256	20.694	= 16	17.256	22.335
= 17	17.970	19.616	= 17	18.835	21.253	= 17	18.835	22.532

λ_k	Lower bounds	Upper bounds	λ_k	Lower bounds	Upper bounds	λ_k	Lower bounds	Upper bounds
$k = 18$	19.573	21.366	$k = 18$	19.573	23.891	$k = 18$	19.573	26.319
$= 19$	19.573	21.366	$= 19$	19.573	23.891	$= 19$	19.573	26.319
$= 20$	20.215	23.975	$= 20$	20.349	26.366	$= 20$	20.349	27.410
$= 21$	21.934	25.963	$= 21$	21.934	26.366	$= 21$	21.855	27.410
$= 22$	22.405	25.963	$= 22$	22.405	28.553	$= 22$	21.952	29.231
$= 23$	22.922	27.095	$= 23$	22.922	28.945	$= 23$	22.922	32.226
$= 24$	22.922	27.095	$= 24$	22.922	29.741	$= 24$	22.922	32.226
$= 25$	23.691	28.165	$= 25$	23.691	29.741	$= 25$	23.691	32.277
$= 26$	23.691	28.498	$= 26$	23.691	30.132	$= 26$	23.691	32.455
$= 27$	25	29.787	$= 27$	25	31.351	$= 27$	25	32.858
$= 28$	25	32.010						
$= 29$	25.232	34.311						
$= 30$	27.018	34.916						
$= 31$	28.400	34.916						
$= 32$	28.400	35.604						
$= 33$	29	37.097						
$= 34$	29	37.171						
$= 35$	30.137	37.380						
$= 36$	30.137	38.306						
$= 37$	30.530	38.306						

$\sigma = 3$

λ_k	Lower bounds	Upper bounds	λ_k	Lower bounds	Upper bounds	λ_k	Lower bounds	Upper bounds
$k = 1$	4.381	4.568	$k = 8$	11.931	16.768	$k = 15$	17.256	23.611
$= 2$	4.381	4.568	$= 9$	11.931	16.768	$= 16$	17.256	23.611
$= 3$	5.136	5.232	$= 10$	13.971	16.825	$= 17$	18.835	25.180
$= 4$	8.395	9.256	$= 11$	14.324	17.355	$= 18$	19.573	27.319
$= 5$	9.623	10.332	$= 12$	14.970	17.855	$= 19$	19.573	27.319
$= 6$	9.623	10.332	$= 13$	15.410	18.160			
$= 7$	10.278	11.376	$= 14$	17.200	23.488			

23. Invariant Subspaces for the Elasticity Operator in a Cube

We want to obtain results analogous to those of the former Section, when the operator under consideration is always the elasticity one

$$Lw \equiv \Delta_2 w + \sigma \text{ grad div } w,$$

considered, now, in the three-dimensional space. Hence w is a real three-component vector; A is the cube $|x_k| < a$ $(k = 1, 2, 3)$. One has first to decompose the space V of the three-component vectors, defined in A, into subspaces which are invariant for the operator L.

This question appears to be somewhat complicated.

Through a suitable analysis, Bassotti [115] shows that the space V of the three-component vectors defined in the cube A is a direct sum of sixteen subspaces invariant with respect to the operator L. The projection operator onto any one of these subspaces is also constructed. Let us briefly describe here the principal result achieved by Bassotti. Let K be the vector space of the real functions defined in A. Let $K^{\alpha_1\alpha_2\alpha_3}$ $(\alpha_i = 0, 1; i = 1, 2, 3)$ be the subspace of K formed by the even (odd) functions of x_i if α_i is 0 (if α_i is 1). K may be decomposed as a direct sum of 8 subspaces in the following way:

$$K = K^{000} \oplus K^{001} \oplus K^{010} \oplus K^{100} \oplus K^{110} \oplus K^{011} \oplus K^{101} \oplus K^{111}.$$

The component $f^{\alpha_1\alpha_2\alpha_3}(x)$ of the real function $f(x) \equiv f(x_1, x_2, x_3)$ in the subspace $K^{\alpha_1\alpha_2\alpha_3}$ is given by:

$$\begin{aligned} f^{\alpha_1\alpha_2\alpha_3}(x_1, x_2, x_3) &= 2^{-3}[f(x_1, x_2, x_3) + (-1)^{\alpha_1} f(-x_1, x_2, x_3) \\ &+ (-1)^{\alpha_2} f(x_1, -x_2, x_3) + (-1)^{\alpha_3} f(x_1, x_2, -x_3) \\ &+ (-1)^{\alpha_1+\alpha_2} f(-x_1, -x_2, x_3) + (-1)^{\alpha_1+\alpha_2} f(-x_1, x_2, -x_3) \\ &+ (-1)^{\alpha_2+\alpha_3} f(x_1, -x_2, -x_3) + (-1)^{\alpha_1+\alpha_2+\alpha_3} f(-x_1, -x_2, -x_3)]. \end{aligned}$$

Let $\Gamma^{\alpha_1\alpha_2\alpha_3}$ be the subspace of V formed by all three-component vector functions $v \equiv (v_1, v_2, v_3)$ such that:

$$v_j \in K^{\beta_1\beta_2\beta_3} \qquad (j = 1, 2, 3)$$

with $\beta_i = \alpha_i + (-1)^{\alpha_i} \delta_i^j (i = 1, 2, 3)$, $[\delta_i^j = 0$ if $i \neq j$, $\delta_i^j = 1$ if $i = j]$.

Let $P^{\alpha_1\alpha_2\alpha_3}$ be the operator projecting V onto $\Gamma^{\alpha_1\alpha_2\alpha_3}$, i.e. the operator defined by the condition

$$P^{\alpha_1\alpha_2\alpha_3} v = v^{\alpha_1\alpha_2\alpha_3} \qquad (v^{\alpha_1\alpha_2\alpha_3} \in \Gamma^{\alpha_1\alpha_2\alpha_3}).$$

For any vector $v \equiv [v_1(x_1, x_2, x_3), v_2(x_1, x_2, x_3), v_3(x_1, x_2, x_3)]$ let us consider the three linear operators $S^{(1)}, S^{(2)}, S^{(3)}$, defined as follows:

$$\begin{aligned} S^{(1)} v &\equiv [v_1(x_1, x_3, x_2), v_3(x_1, x_3, x_2), v_2(x_1, x_3, x_2)], \\ S^{(2)} v &\equiv [v_3(x_3, x_2, x_1), v_2(x_3, x_2, x_1), v_1(x_3, x_2, x_1)], \\ S^{(3)} v &\equiv [v_2(x_2, x_1, x_3), v_1(x_2, x_1, x_3), v_3(x_2, x_1, x_3)]. \end{aligned}$$

Let I be the identity operator. Set:

$$P^{\alpha_1\alpha_2\alpha_3 0} = \frac{1}{2}(I + S^{(i)}) P^{\alpha_1\alpha_2\alpha_3}, \qquad P^{\alpha_1\alpha_2\alpha_3 1} = \frac{1}{2}(I - S^{(i)}) P^{\alpha_1\alpha_2\alpha_3}.$$

Let $\Gamma^{\alpha_1\alpha_2\alpha_3\alpha_4}$ be the range of the operator $P^{\alpha_1\alpha_2\alpha_3\alpha_4}$ ($\alpha_i = 0, 1; i = 1, 2, 3, 4$).

The space V *is a direct sum of the 16 subspaces* $\Gamma^{\alpha_1\alpha_2\alpha_3\alpha_4}$; *each one of these subspaces is invariant for the operator* L.

In the writer's opinion, Bassotti's paper is concerned with investigations of remarkable interest in applied mathematics, and it involves solving the following problem: given a vector differential operator which is invariant with respect to some transformation group (the orthogonal group in the elasticity case) and given a domain which has a prescribed symmetry group, decompose the space of the vectors defined in the domain into subspaces which are invariant with respect to the given differential operator. More recently Bassotti [186] has proved that V may be further decomposed into 20 invariant subspaces. Subsequently Smith [188] has shown that such a decomposition is *maximal*. Both Bassotti [186], [187] and Smith [188] have extended their results to more general domains.

24. Computation of the Orthogonal Invariants for Elliptic Differential Operators: General Theory

Let us consider again the theory of Section 16 in order to achieve the construction of the orthogonal invariants for the elliptic boundary value problems without the restriction imposed in that Section on the problems treated therein.

Consider again, therefore, the self-adjoint operator

$$L(x, D) = \sum_{\substack{0 \leqslant |p| \leqslant m \\ 0 \leqslant |q| \leqslant m}} D^p a_{pq}(x) D^q,$$

the a_{pq} being $l \times l$ complex matrices.

Let $B(u, v)$ be the bilinear form already introduced in Section 16 and V a linear manifold of $H_m(A)$ such that $V \supset \mathring{H}_m(A)$.[23]

Let us assume:

(1) For any pair of multi-indices p and q, let

$$a_{pq}(x) = (-1)^{|p|+|q|} \bar{a}_{qp},$$

(2) $(-1)^m B(w, w) \geqslant c \, \| w \|_m^2 \qquad (c > 0)$ for any $w \in V$.

It follows from strongly elliptic operator theory that, for any fixed $u \in H_0(A)$, there is one and only one element $v \in V$ such that, for any $w \in V$,

$$(-1)^m B(v, w) = (u, w)_0.$$

Set $v = Gu$.

---

(23) $\mathring{H}_m(A)$ is the subspace of $H_m(A)$ of the functions which vanish (in the sense of the functions of $H_m(A)$) together with their derivatives up to the order $m - 1$ on ∂A.

1.XII. *The operator* G, *considered as an operator from* $H_0(A)$ *into* $H_0(A)$, *is positive and compact and belongs to* $\mathfrak{C}^n$ *for any* $n > \frac{r}{2m}$.

The eigenvalue problem (1.2.1) is equivalent to the following one:

$$(-1)^m B(v, w) - \lambda (v, w)_0 = 0 \qquad (v \in V, \forall w \in V) \tag{1.24.1}$$

which gives rise to the differential problem

$$(-1)^m L(x, D) v - \lambda v = 0$$

with further conditions on v (of the 'boundary conditions' type) only if $a_{pq}(x)$'s are smooth enough.

We will consider the theory for the general problem (1.24.1). Let us set

$$L_0(D) = \alpha_{pq} D^{p+q},$$

the α_{pq}'s being $l \times l$ constant matrices such that

$$\alpha_{pq} = (-1)^{|p|+|q|} \bar{\alpha}_{qp}.$$

Put

$$B_0(u, v) = \sum_{|p|,|q|}^{0,m} (-1)^{|p|} \int_A (\alpha_{pq} D^q u) D^p v \, dx$$

and suppose that, for $w \in V$

$$(-1)^m B_0(w, w) \geq c_0 \| w \|_m^2 \qquad (c_0 > 0).$$

Let it further be true that

$$(-1)^m B_0(w, w) \geq 0 \qquad (\forall w \in H_m(A)).$$

It can be assumed, for instance, that

$$L_0(D) = \gamma \sum_{|p|}^{0,m} (-1)^{|p|+m} D^p D^p$$

γ being a positive constant.

We are not constrained, however, to such a particular choice of L_0. We can also suppose that, for any non-zero $w \in V$

$$(-1)^m B(w, w) - (-1)^m B_0(w, w) > 0.$$

Let a linear operator R be known, representing $H_0(A)$ into $H_m(A)$ and such that, for any $w \in V$ and any $u \in H_0(A)$

$$(-1)^m B_0(Ru, w) = (u, w)_0.$$

R is a particular 'fundamental solution' for the L_0 operator. The explicit construction of R may be performed within very general assumptions. For instance, if A has a boundary sufficiently smooth and is homeomorphic to a circular domain, the construction of R is obtained solving a 'Neumann type' problem for a particular constant coefficients operator, in a circular domain.

Let K be the kernel of the quadratic form $B_0(v, v)$ in the space $H_m(A)$. Considering the quotient space $\mathcal{H} = H_m(A)/K$, let us introduce in $\mathcal{H}$ the following new scalar product:

$$[u, v] = (-1)^m B_0(u, v).$$

Since V may be considered as a subspace of $\mathcal{H}$, consider the orthogonal complement Ω of V with respect to $\mathcal{H}$.

That is, let

$$\mathcal{H} = V \oplus \Omega.$$

Define P as the orthogonal projector of the space $\mathcal{H}$ onto Ω and let R^* be the operator from $\mathcal{H}$ into $H_0(A)$ determined by the condition

$$[Ru, w] = (u, R^* w)_0 \qquad [\forall u \in H_0(A),\ \forall w \in \mathcal{H}].$$

Now let $\{\psi_j\}$ be a system of linearly independent vectors, complete in the space $H_0(A)$. Set

$$\varphi_k = G\psi_k - R^*R\psi_k + R^*PR\psi_k.$$

One proves [73], [74], [75] that the matrix

$$\{(\psi_j, \varphi_k)_0\} \qquad (j, k = 1, \ldots, \nu)$$

is non-singular. Let $\{\sigma_{jk}^{(\nu)}\}$ be its inverse matrix. Set

$$Q_\nu u = \sigma_{jk}^{(\nu)} (u, \varphi_k)_0 \varphi_j.$$

Now, *the operator* Q_ν *converges to a continuous linear operator* Q *in the strong topology of the continuous linear operators on* $H_0(A)$.

The following structure theorem for the operator G holds [73].

1.XIII. *The operator* G *has the following representation*:

$$G = R^*R - R^*PR - Q. \tag{1.24.2}$$

Let $\{\omega_s\}$ be a system of linearly independent vectors, complete in Ω and let $P^{(\rho)}$ be the orthogonal projector of $\mathcal{H}$ onto the manifold spanned by $\omega_1, \ldots, \omega_\rho$.

The following theorem holds:

1.XIV. *Setting*:

$$\Gamma_\nu^{(\rho)} = R^*R - R^*P^{(\rho)}R - Q_\nu,$$

one has:

$$\mathfrak{z}_s^n(\Gamma_\nu^{(\rho)}) \geqslant \mathfrak{z}_s^n(\Gamma_{\nu'}^{(\rho')}) \qquad \text{if } \rho \leqslant \rho', \qquad \nu \leqslant \nu'$$

and furthermore:

$$\lim_{\substack{\rho\to\infty \\ \nu\to\infty}} \mathfrak{z}_s^n(\Gamma_\nu^{(\rho)}) = \mathfrak{z}_s^n(\mathrm{G}).$$

Since $\mathfrak{z}_s^n(\Gamma_\nu^{(\rho)})$ is explicitly computable (*see* (1.10.2)), this theorem yields the possibility of obtaining upper approximations on $\mathfrak{z}_s^n(\mathrm{G})$ and, hence, of constructing lower bounds to λ_k.

25. The Eigenfrequencies of a Heterogeneous Elastic Body

The theory reviewed in the former Section adapts itself to many different applications. Among these applications let us consider the computation of the eigenfrequencies of an elastic body formed by two different anisotropic and heterogeneous elastic bodies, when one is clamped onto the other. Such an application has been developed in [77].

Let A be a regular domain, supposed to be a subset of the space X^r, in order to include in a unified treatment, both cases of physical interest $r = 2$ and $r = 3$. Let A be decomposed into two domains A_1 and A_2, disjoint and separated by a regular hypersurface Σ. Let A_1 and A_2 be filled by two elastic bodies in their natural configurations.

Let $W^{(k)}(x, \epsilon)$ be the elastic potential relative to the body filling A_k ($k = 1, 2$):

$$W^{(k)}(x, \epsilon) = \frac{1}{2} \sum_{ihjl}^{1,r} a_{ih,jl}^{(k)}(x)\, \epsilon_{ih}\, \epsilon_{jl}.$$

$\epsilon = \{\epsilon_{ih}\}$ stands for the strain tensor, whose components are defined, through those of the displacement vector u, in the following way:

$$\epsilon_{ih} \equiv \epsilon_{ih}(u) = 2^{-1}(u_{i/h} + u_{h/i}).$$

The functions $a_{ih,jl}^{(k)}(x)$ are real and of class C^∞ in $\bar{A}_k$ and the quadratic form $W^{(k)}(x, \epsilon)$ is positive definite in the variables ϵ_{ih}, with $\epsilon_{ih} = \epsilon_{hi}$.

Moreover let

$$a_{ih,jl}^{(k)}(x) \equiv a_{jl,ih}^{(k)}(x); \qquad a_{ih,jl}^{(k)}(x) \equiv a_{hi,jl}^{(k)}(x).$$

Decompose ∂A into two disjoint sets $\partial_1 A$ and $\partial_2 A$, which will be assumed to be formed by pieces of open regular hypersurfaces, in such a way that

$$\partial A = \overline{\partial_1 A} \cup \overline{\partial_2 A}.$$

Consider the following problem (P), which corresponds to the problem of the equilibrium of a body A, kept fixed along $\partial_1 A$, free along $\partial_2 A$ and subject to an assigned system of body forces.

(P): *Given the r-component vector functions $f^{(1)}$ and $f^{(2)}$ in A_1 and A_2, respectively,* *find the vector function u, which satisfies the following conditions:*

(a) $$\sum_{h=1}^{r} \frac{\partial}{\partial x_h} \frac{\partial}{\partial \epsilon_{ih}} W^{(k)}(x, \epsilon) + f^{(k)} = 0 \qquad \text{in } A_k \qquad (i = 1, \ldots, r; k = 1, 2).$$

(b) *If* $\bar{\Sigma} \cup \bar{\Sigma}_k$ *is the boundary of* A_k, *let*

$$u = 0 \qquad \text{on } \partial_1 A, \qquad \sum_{h=1}^{r} \nu_h \frac{\partial}{\partial \epsilon_{ih}} W^{(k)}(x, \epsilon) = 0 \qquad \text{on } \partial_2 A \cap \Sigma_k$$

(ν *is the inward normal in the points of* $\partial_2 A$).

(c) u *is continuous in* A.

(d) *If* $\nu^{(k)}$ *is the normal to* Σ, *directed towards the interior of* $A_k (k = 1, 2)$ (Fig. 1.6), *then*

$$\sum_{h=1}^{r} \left(\nu_h^{(1)} \frac{\partial W^{(1)}}{\partial \epsilon_{ih}} + \nu_h^{(2)} \frac{\partial W^{(2)}}{\partial \epsilon_{ih}} \right) = 0 \qquad (i = 1, \ldots, r).$$

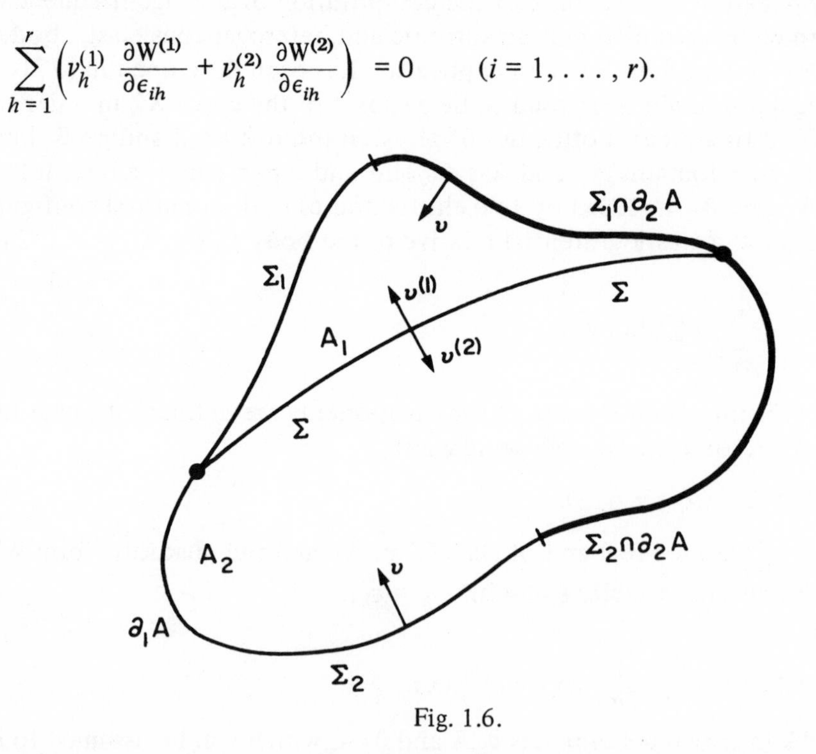

Fig. 1.6.

Problems of this type were first considered by Picone [116], when A is composed by two homogeneous bodies A_1 and A_2, having different Lamé constants. The very same case was later studied by Lions [117] and Campanato [118]. The general case has been considered in [102], where an existence theorem is given for a smooth solution, if such are the data of the problem. Subsequently the corresponding eigenvalue problem has been studied [77]. It comprises the differential system

$$\sum_{h=1}^{r} \frac{\partial}{\partial x_h} \frac{\partial}{\partial \epsilon_{ih}} W^{(k)}(x, \epsilon) + \lambda u_i = 0 \qquad \text{in } A_k \qquad (i = 1, \ldots, r; k = 1, 2),$$

together with the imposition of conditions (b), (c), (d), specified above on the unknown vector u.

One first proves that the problem admits a denumerable infinity of positive eigenvalues

$$0 < \lambda_1 \leqslant \lambda_2 \leqslant \ldots \leqslant \lambda_n \leqslant \ldots$$

with $\lim_{k \to \infty} \lambda_k = +\infty$.

The upper bounds to the λ_k's are computed through a method that extends the Rayleigh–Ritz one to the problem under consideration.

Setting

$$B(u, v) = \sum_{k=1}^{2} \sum_{ihjl}^{1,r} \int_{A_k} a_{ih,jl}^{(k)} \epsilon_{ih}(u)\, \epsilon_{jl}(v)\, dx,$$

the upper approximations to the first ν eigenvalues are obtained as the roots of the equation

$$\det \{B(v_i, v_j) - \lambda(v_i, v_j)\} = 0 \qquad (i, j = 1, \ldots, \nu)$$

$\{v_i\}$ being a system of (r-vector) functions, vanishing on $\partial_1 A$ and satisfying suitable completeness conditions.

To construct lower approximations assume the existence of a bi-lipschitzian homeomorphism $\xi = \kappa(x)$ mapping the closure $\bar{A}$ of A onto the closure $\bar{D}$ of a domain D of the space of the variables $\xi_1, \ldots, \xi_r$, in such a way that $\partial_2 A$ is mapped in a $(r-1)$-dimensional domain of the space $\xi_r = 0$ and the image of $\bar{A} - \overline{\partial_2 A}$ is contained in the half-space $\xi_r > 0$ (Fig. 1.7).

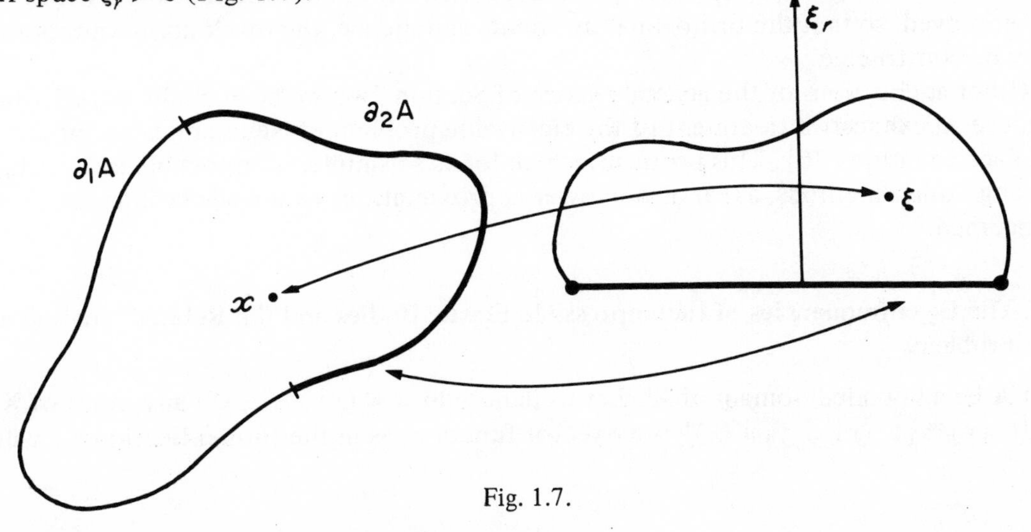

Fig. 1.7.

Remembering the definition of B (u, v), one has

$$B(v, v) = \sum_{ihjl} \int_D \alpha_{ihjl}(\xi)\, w_{i/h}\, w_{j/l} d\xi \equiv \mathcal{B}(w, w),$$

where the α_{ihjl}'s are bounded and measurable functions on D and $w(\xi) = v[\kappa^{-1}(\xi)]$.

There exists a $\sigma_0 > 0$ such that

$$\mathcal{B}(w, w) \geqslant \sigma_0 \sum_{ih} \int_D (w_{i/h})^2 d\xi$$

Let $s(\xi - \eta)$ be defined as $s(x, y)$ on page 70. Set:

$$Rf = \frac{1}{\sigma_0} \int_A s[\kappa(x) - \kappa(y)]\, f(y)\, dy$$

$$\mathcal{B}_0(w, z) = \sum_{ih} \int_D w_{i/h} z_{i/h}\, d\xi$$

$$B_0(u, v) = \mathcal{B}_0(u[\kappa^{-1}(\xi)], v[\kappa^{-1}(\xi)]).$$

For any $v \in H_1(A)$, vanishing on $\partial_1 A$, one has:

$$B_0(Rf, v) = \int_A fv\, dx.$$

The form B_0 and the operator R satisfy the conditions required in the former Section, in order to apply the theory developed therein. Therefore that theory may be employed, so that the orthogonal invariants and, hence, the lower approximations may be constructed.

Other applications of the general theory of Section 24 have been made, in order to achieve an exhaustive treatment of the eigenvalue problem of Neumann type for the Laplace equation [76]. This problem, which looks so simple, is impossible to investigate through other methods, as far as the lower approximations to the eigenvalues are concerned.

26. The Eigenfrequencies of Incompressible Elastic Bodies and the Related Analytical Problems

Let A be a bounded domain of X^r. Let us denote by $x \equiv (x^1, \ldots, x^r)$ any point of X^r.

If $v(x) \equiv \{v^1(x), \ldots, v^r(x)\}$ is a r-vector function, as in the former Section we will set:

$$\epsilon_{ik}(v) = \frac{1}{2}\left(\frac{\partial v^i}{\partial x^k} + \frac{\partial v^k}{\partial x^i}\right).$$

Let

$$W(\epsilon) = \epsilon_{ik}\epsilon_{ik} + \sigma\epsilon_{ii}\epsilon_{kk}.$$ [24]

It is well known that if A (for $r = 2$ or $r = 3$) represents the natural configuration of a two- or three-dimensional isotropic and homogeneous elastic body, then $W(\epsilon)$ is (up to an inessential constant factor) the elastic potential of the body; σ is a constant depending on the elastic properties of the material.

Let the body be fixed along the boundary and subject to a given system of body forces, expressed by the r-vector function $f(x)$. It is well known that the *equilibrium configuration* is obtained by minimizing the *energy functional*

$$I[v] = \int_A W(\epsilon)\, dx - \int_A vf\, dx$$

within the class of the admissible displacements vanishing on ∂A.

The problem of vibration of the body, represented by A, consists in considering the complete continuity of the quadratic form

$$\Psi[v] = \int_A |v|^2\, dx$$

with respect to the quadratic form

$$\Phi[v] = \int_A W(\epsilon)\, dx.$$

More precisely, the eigenfrequencies of the body are expressed by means of the eigenvalues of $\Psi[v]$ with respect to $\Phi[v]$.

We have already solved this problem in Section 17. We want to consider, now, the problem of minimizing $I[v]$ (static problem) or that of studying the complete continuity of $\Psi[v]$ with respect to $\Phi[v]$ (vibratory problem), within the further hypothesis (*incompressibility constraint*) that

$$\epsilon_{ii} = 0 \qquad \text{in A.} \tag{1.26.1}$$

It is easily seen that, if v is an admissible function satisfying (1.26.1) and such that

$$v = 0 \qquad \text{on } \partial A, \tag{1.26.2}$$

one has:

---

[24] In this Section use will be made of the convention that a repeated index means a summation over that index.

$$\Phi[v] = \frac{1}{2}\int_A \sum_{i=1}^{r} |\operatorname{grad} v^i|^2 \, dx. \tag{1.26.3}$$

We shall consider the following problems:

(1) (*Static problem*). Minimize the functional

$$I[v] = \frac{1}{2}\int_A \sum_{i=1}^{r} |\operatorname{grad} v^i|^2 \, dx - \int_A vf \, dx \tag{1.26.4}$$

within the class of the admissible displacements satisfying (1.26.1) and (1.26.2).

(2) (*Vibratory problem*). Consider the complete continuity of the quadratic form $\Psi[v]$ with respect to $\Phi[v]$, defined by (1.26.3), within the class of the admissible displacements satisfying (1.26.1) and (1.26.2).

For $r = 2$ problem (1) may be reformulated in this way: minimize the functional (1.26.4) within the class of the functions v^1, v^2, such that

$$\frac{\partial v^2}{\partial x^1} - \frac{\partial v^1}{\partial x^2} = 0 \quad \text{in A}, \qquad v^1 = v^2 = 0 \quad \text{on } \partial A.$$

If A is supposed to be simply connected, then problem (1) consists in minimizing the functional

$$J(u) = \frac{1}{2}\int_A (u^2_{x^1x^1} + 2u^2_{x^1x^2} + u^2_{x^2x^2}) \, dx + \int_A u\,(f^1_{x^1} + f^2_{x^2}) \, dx$$

within a class of admissible scalar functions, whose gradient vanishes on ∂A.

We are, thus, led to the classical variational problem connected with the biharmonic problem

$$\Delta_2\Delta_2 u = -(f^1_{x^1} + f^2_{x^2}) \quad \text{in A} \tag{1.26.5}$$

$$u_{x^1} = u_{x^2} = 0 \quad \text{on } \partial A.$$

Similar arguments show that problem (2) is equivalent to the classical eigenvalue problem connected with the buckling of a clamped plate

$$\Delta_2\Delta_2 u + \lambda\Delta_2 u = 0 \quad \text{in A} \tag{1.26.6}$$

$$u = u_{x^1} = u_{x^2} = 0 \quad \text{on } \partial A.$$

The equivalence between problem (1) and (1.26.5), as well as that between problem (2) and (1.26.6), ceases to be true for $r > 2$.

However, generalizing not in the trivial way (1.26.5) and (1.26.6) to $r > 2$, one always gets equivalence between problems (1) and (2) and certain boundary value problems which reduce to the classical ones, now considered, for $r = 2$.

This matter is part of a theory developed by Colautti [119] in an even wider context, with respect to the requirements of the physical problem, but of remarkable mathematical interest.

The principal mathematical objects of this theory are the exterior differential forms in the sense of Cartan. An *exterior differential form of degree* k (briefly, a k*-form*) is represented, in the coordinate system $x^1, \dots, x^r$. in the following way:

$$u = \frac{1}{k!} u_{s_1 \dots s_k} \mathrm{d}x^{s_1} \cdots \mathrm{d}x^{s_k}.$$

The coefficients $u_{s_1 \dots s_k}$ of the form are the components of a real, antisymmetric and covariant tensor.

For $k = 0$, u is a real function and, for $k > r$, $u \equiv 0$.

If v is a k-form

$$v = \frac{1}{h!} v_{i_1 \dots i_h} \mathrm{d}x^{i_1} \cdots \mathrm{d}x^{i_h},$$

the exterior product $u \wedge v$ is the $(h + k)$-form

$$u \wedge v = \frac{1}{k!h!} u_{s_1 \dots s_k} v_{i_1 \dots i_h} \mathrm{d}x^{s_1} \cdots \mathrm{d}x^{s_k} \mathrm{d}x^{i_1} \cdots \mathrm{d}x^{i_h}.$$

The *adjoint form* of the form u is the $(r - k)$-form $* u$, defined in the following way:

$$* u = \frac{1}{(r-k)!} \overset{*}{u}_{i_1 \dots i_{r-k}} \mathrm{d}x^{i_1} \cdots \mathrm{d}x^{i_{r-k}},$$

where

$$\overset{*}{u}_{i_1 \dots i_{r-k}} = \frac{1}{k!} \delta^{1 \dots \dots \dots \dots r}_{s_1 \dots s_k i_1 \dots i_{r-k}} u_{s_1 \dots s_k}.$$

One has: $* * u = (-1)^{rk+k} u$.

If the k-form u has differentiable coefficients, the *differential of* u, denoted by $\mathrm{d}u$, is defined to be the $(k + 1)$-form

$$\mathrm{d}u = \frac{1}{k!} \mathrm{d}u_{s_1 \dots s_k} \mathrm{d}x^{s_1} \cdots \mathrm{d}x^{s_k}$$

$$= \frac{1}{(k+1)!} \left(\sum_{h=1}^{k+1} (-1)^{h-1} \frac{\partial u_{s_1 \dots s_{h-1} s_{h+1} \dots s_{k+1}}}{\partial x^{s_h}} \right) \mathrm{d}x^{s_1} \cdots \mathrm{d}x^{s_{k+1}}.$$

One has (theorem of Volterra–Poincaré)

$$\mathrm{dd}u = 0.$$

The *co-differential* δu of the k-form is the $(k - 1)$-form given by

$$\delta u = (-1)^{r(k+1)+1} * \mathrm{d} * u.$$

It is obvious that $\delta \delta u = 0$.

The following operator: $\Delta = -\delta d$ is called the *Volterra harmonic operator*, whereas the operator $\Delta_2 = -(\delta d + d\delta)$ is called the *Laplace harmonic operator*. For $k = 0$ one has:

$$\Delta \equiv \Delta_2 \equiv \sum_{i=1}^{r} \frac{\partial^2}{\partial x_i^2}.$$

$H_m^k(A)$ is the space of the k-forms, whose coefficients belong to $H_m(A)$. If u and v are k-forms of $H_0^k(A)$, we shall set:

$$(u, v) = \int_A u \wedge *v, \qquad \| u \| = (u, u)^{1/2}.$$

Consider the following two general problems:

(VS) *Variational static problem*: Let f be a $(k+1)$-form $(0 \leqslant k < r-1)$ of $H_1^{k+1}(A)$ and let $V^{k+1}(A)$ be the space of the $(k+1)$-forms defined by the following conditions:

(a) $v \in H_1^{k+1}(A)$, (b) $v_{|\partial A} = 0$,[25] (c) $dv = 0$ in A.

Minimize the functional

$$I[v] = \frac{1}{2} \| \delta v \|^2 - (v, f)$$

within the class $V^{k+1}(A)$.

(VV) *Variational vibratory problem*. Setting $\Psi(v) = \| v \|^2$, $\Phi(v) = \| \delta v \|^2$, prove the complete continuity of $\Psi(v)$ with respect to $\Phi(v)$ and compute its eigenvalues.

Problems (1) and (2) mentioned above are easily proved to be particular cases of (VS) and (VV), respectively, for $k = r - 2$.

Denote by $Q_m^k(A)$ the intersection of all spaces $H_m^k(B)$ for any B such that $\bar{B} \subset A$.

Consider the following problem:

(BV) *Biharmonic Volterra problem*. Let $f \in H_1^{k+1}(A)$ $(0 \leqslant k < r-1)$. Find a k-form satisfying the following conditions:

(a) $u \in H_2^k(A) \cap Q_4^k(A) \equiv \mathfrak{A}_{2,4}^k(A)$,

(b) $\Delta\Delta f = \delta f$ in A, $du_{|\partial A} = 0$.

In the particular case $k = 0$, $r = 2$ problem (BV) reduces to the classical biharmonic problem (1.26.5).

Finally we will call the *generalized buckling problem* the following one:

(GB) Consider, within the class $\mathfrak{A}_{2,4}^k(A)$, the eigenvalue problem:

$$\Delta\Delta u + \lambda\Delta u = 0 \quad \text{in A}, \qquad du_{|\partial A} = 0.$$

(25) When writing $v_{|\partial A}$ we mean the $(k+1)$-form of X^r, defined on ∂A, whose coefficients are the traces on ∂A of the coefficients of v. Hence $v_{|\partial A} = 0$ means that all coefficients of v vanish on ∂A.

As an *eigensolution* for this eigenvalue problem must be understood to be any k-form u which satisfies the above differential equation and the above boundary condition and which is *not* closed, i.e. is such that du does *not* vanish identically in A.

The following hypotheses on A are made: (1) A is properly regular; (2) $\bar{A}$ is C^∞-homeomorphic to the closure $\bar{D}$ of a starlike domain.

The following theorems hold.

1.XV. *Problem* (VS) *is equivalent to problem* (BV) *in the sense that* v *minimizes* $I[v]$ *in* $V^{k+1}(A)$, *if and only if* $v = du$, u *being a solution of* (BV).

1.XVI. *Problem* (VV) *is equivalent to problem* (GB) *in the sense that the eigenvalues of* (VV) *are eigenvalues of* (GB) *and conversely. Any eigenvalue has the same multiplicity in both cases. Moreover, if* $\{u_n\}$ *is a complete system of eigensolutions of* (GB), *then* $\{du_n\}$ *is a complete system of eigensolutions of* (VV).

It is worth remarking that (BV) is equivalent to the following problem, whose unknown is the $(k + 1)$-form v:

$$d\delta v + \delta\varphi = f, \qquad dv = 0 \qquad \text{in A},$$

$$v|_{\partial A} = 0.$$

If, for $r = 3$, $k = 1$, the vector notations are used instead of the ones relative to exterior forms such a problem becomes

$$\Delta_2 \boldsymbol{v} + \operatorname{grad} p + f = 0, \qquad \operatorname{div} \boldsymbol{v} = 0 \qquad \text{in A}, \tag{1.26.7}$$

$$\boldsymbol{v}|_{\partial A} = 0,$$

$\boldsymbol{v}$ being a three-vector function and p a scalar real function. The problem is the classical linearized Navier–Stokes one of the viscous fluid theory.

Let us briefly mention the way to achieve the solutions of problems (BV) and (GB).

Consider the *Hodge parametrix* [120]

$$\omega_k(x, y) = \sum_{s_1 < \ldots < s_k} s(x, y)\, dx^{s_1} \cdots dx^{s_k}\, dy^{s_1} \cdots dy^{s_k},$$

where $s(x, y)$ is defined as on page 70. Introduce the two operators

$$Tv = \int_A \omega_k(x, y) \wedge * v(y),$$

$$Rv = \int_A d_y\, \omega_k(x, y) \wedge * v(y).$$

T maps k-forms into k forms and R $(k + 1)$-forms into k-forms.

Let Q be the linear manifold of the k-forms h $(0 \leqslant k \leqslant r)$ satisfying the following condition: for any h there exist a domain B, such that $B \supset \bar{A}$, and a $(k+1)$-form a, such that $(d\delta + \delta d)\, a = 0$, in such a way that in B: $h = \delta a$. Let P be the orthogonal projector of $H_0^k(A)$ onto the closure $\bar{Q}$ of Q (the 'closure' must be understood in the topology of $H_0^k(A)$).

1.XVII. *The most general solution of problem* (BV) *is given by*

$$u = T(R - PR)\, f + u_0,$$

u_0 *being an arbitrary closed* (*i.e. such that* $du_0 = 0$) *k-form of* $\mathfrak{A}^k_{2,4}(A)$.

The operator $T(R - PR)$ is, thus, the Green operator of problem (BV).

The construction of this operator, for which the compactness properties required by general eigenvalue theory are proved, allows us to prove the following theorem.

1.XVIII. *Let* $W^{k+1}(A)$ *be the closure of* $V^{k+1}(A)$ *in the topology of* $H_0^k(A)$. *Problem* (VV) *is equivalent to the eigenvalue problem*

$$dT(R - PR) - \mu v = 0 \qquad v \in W^{k+1}(A) \tag{1.26.8}$$

in the sense that μ *is an eigenvalue of* (1.26.8) *of multiplicity m, if and only if* μ^{-1} *is an eigenvalue of* (VV) (*and hence of* (GB)), *with the same multiplicity.*

The orthogonal invariants theory may be applied to the operator $G = dT(R - PR)$ since it is the uniform limit of the decreasing operator sequence $G_\rho = dT(R - P_\rho R)$ (P_ρ having its usual meaning).

The theory just developed allows us, in particular, to solve the static and the vibratory problems for the linearized Navier–Stokes system (1.26.7).

In [121] Colautti considered the case when A is an elastic, incompressible three-dimensional sphere, clamped along the boundary: axially symmetric vibrations are considered.

Then one has, according to the general theory, the eigenvalue problem (GB), which, if vector notations are used instead of the exterior form ones, may be written out in the following way:

$$\operatorname{rot}\,[\operatorname{rot}\operatorname{rot} \mathbf{v} - \lambda v] = 0; \qquad \operatorname{div} v = 0 \qquad \text{in A}$$

$$\mathbf{v} = 0 \qquad \text{on } \partial A.$$

In addition, denoting by $v_\rho, v_\varphi, v_\theta$ the polar components of the vector v, by the axial symmetry hypothesis it must be

$$\frac{\partial \mathbf{v}_\rho}{\partial \theta} = 0, \qquad \frac{\partial \mathbf{v}_\rho}{\partial \theta} = 0, \qquad v_\theta = 0 \qquad \text{in A.}$$

In this particular case, indicating by $u = u(\rho)$ a function of ρ, the eigenvalue problem is equivalent to the following sequence of eigenvalue problems in the interval (0,1)

$$L_n L_n u + \lambda L_n u = 0 \qquad (n = 1, 2, \dots)$$

$$u(1) = u'(1) = 0, \qquad \int_0^1 |L_n u|^2 \, d\rho < +\infty \qquad (n = 1, 2, \dots) \qquad (1.26.9)_n$$

where

$$L_n u = \frac{d^2 u}{d\rho^2} - n(n+1)\frac{u}{\rho^2}$$

Assuming $\mu = \lambda^{-1}$, problem $(1.26.9)_n$ is equivalent to the following one:

$$G_n f - \mu f = 0, \qquad f \in \mathfrak{L}^2(0, 1),$$

where

$$G_n f = \left[-\frac{2n+3}{2(2n+1)} \int_0^1 r^{n+1} f(r)\, dr + \frac{1}{2} \int_0^1 r^{n+3} f(r)\, dr \right] \rho^{n+1}$$

$$+ \frac{1}{2n+1} \left[\int_0^\rho \rho^{-n} r^{n+1} f(r)\, dr + \int_\rho^1 \rho^{n+1} r^{-n} f(r)\, dr \right]$$

$$- \frac{2n+3}{2(2n+5)} \rho^{n+1} \int_0^1 r^{n+1} f(r)\, dr + \frac{1}{2} \rho^{n+3} \int_0^1 r^{n+1} f(r)\, dr.$$

The upper bounds have been obtained through the Rayleigh–Ritz method, using the following functions:

$$u_h(\rho) = \rho^{2+h}(\rho - 1)^2 \qquad (h = 1, 2, \dots).$$

To obtain lower bounds $\mathfrak{z}_1^4(G_n)$ has been computed. It is given by

$$\mathfrak{z}_1^4(G_n) = \frac{(2n+3)^2 (n+4)^2}{(2n+5)^4 (2n+7)^2 (2n-1)^2} + \frac{1}{16(2n+7)^2 (2n+3)^2}$$

$$+ \frac{(2n+3)^2 (n+3)}{(2n+5)^2 (2n+1)^2 (2n+7)} + \frac{(2n+3)(n+2)}{(2n+5)^4 (2n+1)^2}$$

$$+ \frac{1}{32(2n+7)^2 (2n+5)(2n+3)} + \frac{2n+9}{32(2n+7)^2 (2n+5)^2 (2n+11)}$$

$$+ \frac{1}{16(4n^2-1)^2 (2n+3)} + \frac{1}{16(2n+3)^2 (2n+1)^2 (2n+7)}$$

$$+ \frac{(2n+3)(n+4)}{2(2n+5)^4 (2n+7)(2n-1)} + \frac{(2n+3)(n+4)}{2(2n+5)^3 (2n+7)^2 (2n-1)}$$

$$+\frac{2(2n+3)(n+4)}{(2n+5)^3(2n+7)^2(4n^2-1)}+\frac{1}{8(2n+5)^2(2n+9)(2n+3)}$$

$$+\frac{1}{2(2n+3)^2(2n+11)(2n+7)(2n+1)}+\frac{4(n+2)}{(2n+5)^2(2n+9)(2n+1)(4n^2-1)}$$

$$+\frac{2n+7}{2(2n+5)^2(2n+11)(2n+9)(2n+3)(2n+1)}-\frac{2(2n+3)^2(n+4)}{(2n+5)^4(2n+7)(4n^2-1)}$$

$$-\frac{2(2n+3)(n+4)}{(2n+7)^2(2n+5)^2(2n-1)(4n^2-1)}-\frac{1}{2(2n+5)^2(2n+7)(2n+1)}$$

$$-\frac{1}{2(2n+11)(2n+5)(2n+3)(4n^2-1)}-\frac{n+2}{4(2n+5)^3(2n+1)(n+3)}$$

$$-\frac{(2n+3)(n+4)}{4(2n+5)^2(2n+9)(2n+7)(2n+1)(n+3)}-\frac{4(n+3)}{(2n+5)^2(2n+1)^2(2n+9)(2n+7)}$$

$$-\frac{1}{2(2n+11)(2n+7)(2n+3)(4n^2-1)}-\frac{1}{8(2n+5)(2n+3)(2n+1)(4n^2-1)}.$$

Problems $(1.26.9)_n$ correspond to problems relative to invariant subspaces for the initial problem. They give rise to the following bounds on the relative eigenvalues.

$n = 1$

λ_k	Lower bounds	Upper bounds
$k = 1$	33.217456	33.217462
$k = 2$	82.7186	82.7193
$k = 3$	151.843	151.855
$k = 4$	240.59	240.71
$k = 5$	348.5	349.3
$k = 6$	474.2	477.6
$k = 7$	613.1	625.7
$k = 8$	756.9	795.8
$k = 9$	888.4	987.1
$k = 10$	1019	1280

$n = 2$

λ_k	Lower bounds	Upper bounds
$k = 1$	48.831155	48.831194
$k = 2$	108.5142	108.5164
$k = 3$	187.603	187.636
$k = 4$	286.14	286.41
$k = 5$	403.4	404.9
$k = 6$	536.7	543.1
$k = 7$	679.7	701.5
$k = 8$	820.9	882.4
$k = 9$	951	1107
$k = 10$	1054	1406

$n = 3$

λ_k	Lower bounds	Upper bounds
$k = 1$	66.95413	66.95432
$k = 2$	136.998	137.005
$k = 3$	226.111	226.192
$k = 4$	334.37	334.94
$k = 5$	460.5	463.4
$k = 6$	600	612
$k = 7$	746	782
$k = 8$	881	972
$k = 9$	1014	1254

$n = 4$

λ_k	Lower bounds	Upper bounds
$k = 1$	87.5305	87.5313
$k = 2$	168.113	168.131
$k = 3$	267.30	267.48
$k = 4$	385.15	386.25
$k = 5$	519.6	524.7
$k = 6$	665	683
$k = 7$	810	863
$k = 8$	946	1081
$k = 9$	1058	1368

$n = 5$

λ_k	Lower bounds	Upper bounds
$k = 1$	110.5177	110.5198
$k = 2$	201.810	201.851
$k = 3$	311.10	311.46
$k = 4$	438.35	440.31
$k = 5$	580.5	588.7
$k = 6$	730	758
$k = 7$	873	984
$k = 8$	1013	1207

$n = 6$

λ_k	Lower bounds	Upper bounds
$k = 1$	135.8813	135.8864
$k = 2$	238.041	238.125
$k = 3$	357.44	358.08
$k = 4$	493.8	497.1
$k = 5$	643	655.6
$k = 6$	795	835
$k = 7$	939	1042
$k = 8$	1068	1317

$n = 7$

λ_k	Lower bounds	Upper bounds
$k = 1$	163.592	163.605
$k = 2$	276.76	276.93
$k = 3$	406.24	407.34
$k = 4$	551.44	556.55
$k = 5$	706.9	725.4
$k = 6$	860	915
$k = 7$	1008	1149
$k = 8$	1117	1415

$n = 8$

λ_k	Lower bounds	Upper bounds
$k = 1$	193.626	193.651
$k = 2$	317.93	318.22
$k = 3$	457.4	459.2
$k = 4$	611	618.7
$k = 5$	771.9	797.8
$k = 6$	927	999
$k = 7$	1074	1255
$k = 8$	1174	1538

$n = 9$			$n = 10$			$n = 11$		
λ_k	Lower bounds	Upper bounds	λ_k	Lower bounds	Upper bounds	λ_k	Lower bounds	Upper bounds
$k = 1$	225.960	226.006	$k = 1$	260.57	260.66	$k = 1$	297.44	297.58
$k = 2$	361.50	361.99	$k = 2$	407.4	408.2	$k = 2$	455.69	456.85
$k = 3$	510.9	513.6	$k = 3$	566.5	570.6	$k = 3$	624.4	630.1
$k = 4$	672.5	683.6	$k = 4$	735	751	$k, = 4$	800	821
$k = 5$	837	873	$k = 5$	904	951	$k = 5$	973	1033
$k = 6$	996	1091	$k = 6$	1066	1187	$k = 6$	1134	1282
$k = 7$	1133	1351	$k = 7$	1192	1449	$k = 7$	1257	1563

These bounds are so sharp that they even 'separate' the first 25 computed eigenvalues; therefore, without using Theorem 1.XI, one obtains the following table for these eigenvalues:

λ_k	Lower bounds	Upper bounds
$k = 1$	33.217456	33.217462
$k = 2$	48.831155	48.831194
$k = 3$	66.95413	66.95432
$k = 4$	82.7186	82.7193
$k = 5$	87.5305	87.5313
$k = 6$	108.5142	108.5164
$k = 7$	110.5177	110.5198
$k = 8$	135.8813	135.8864
$k = 9$	136.998	137.005
$k = 10$	151.843	151.855
$k = 11$	163.592	163.605
$k = 12$	168.113	168.131
$k = 13$	187.603	187.636
$k = 14$	193.626	193.651
$k = 15$	201.810	201.851
$k = 16$	225.960	226.006
$k = 17$	226.111	226.192
$k = 18$	238.041	238.125
$k = 19$	240.59	240.71
$k = 20$	260.57	260.66
$k = 21$	267.30	267.48
$k = 22$	276.76	276.93
$k = 23$	286.14	286.41
$k = 24$	297.44	297.58
$k = 25$	311.10	311.46

27. Computation of the Eigenvalue's Multiplicity

Very recently [122] a problem has been considered which seems to be of remarkable interest and difficulty. It involves the *computation of the (geometric) multiplicity of the eigenvalue* μ_k *of the operator* G, considered in Section 2, and, hence, of the multiplicity of the eigenvalue λ_k.

Proceeding in the same way as for the eigenvalues computation it is convenient to look first for the largest transformation group of the Hilbert space S, which leaves the problem of computing the multiplicity invariant. A suitable analysis shows that such a problem is invariant with respect to the group of the roto-homotheties of S, i.e. the group whose most general transformation is obtained as the product between a rotation (unitary transformation of S) and a homothety (centred, of course, on the origin of S).

It will, therefore, be reasonable to require the upper and lower approximations to the multiplicity of each eigenvalue to be provided by some *roto-homothetic invariants* (i.e. the invariants with respect to the roto-homothetic group).

Consider the case $k = 1$. Then, denoting by p the multiplicity of μ_1, one tries to construct two sequences $\{\varphi_\nu(G)\}$ and $\{\psi_\nu(G)\}$ of scalar functions of the operator G, such that any one of these functions is a roto-homothetic invariant and such that, for $\nu \to \infty$, the first sequence converges from below and the second one from above to p.

Let $f(G)$ be a scalar function of the operator G, defined in $\mathcal{C}^n$, which is an orthogonal invariant, i.e. such that

$$f(G) = f(U^{-1}GU),$$

U being an arbitrary unitary operator. We shall say that f is homogeneous of degree α if, for any $t > 0$, one has:

$$f(tG) = t^\alpha f(G).$$

It is immediately verified that *any homogeneous function of degree zero is invariant with respect to the unitary group when, and only when, it is a roto-homothetic invariant.*

Moreover, a complete system of roto-homothetic invariants may be constructed in this way.

One has, then, to try choosing $\varphi_\nu(G)$ and $\psi_\nu(G)$ among functions which are orthogonal invariants homogeneous of degree zero and which satisfy the conditions required above. A possible choice is:

$$\psi_\nu(G) = \mu_1^{-\nu s}\, \mathfrak{I}_s^\nu(G),$$

s being a fixed positive integer.

It is proved that

$$\psi_\nu(G) > \psi_{\nu+1}(G) \tag{1.27.1}$$

and furthermore that

$$\lim_{\nu \to \infty} \psi_\nu(G) \begin{cases} = 0 & \text{if } p < s \\ = \binom{p}{s} & \text{if } p \geqslant s. \end{cases} \tag{1.27.2}$$

Hence, for the case $s = 1$, one has the following result, known in a particular case:[26]

$$\lim_{\nu \to \infty} \psi_\nu(G) = p.$$

(1.27.1) and (1.27.2) allow us to give actual numerical bounds on p. For, if $s > 0$ and $\nu \geqslant n$ are two integer numbers, for which an upper approximation to $\mu_1^{-\nu s}\ \mathfrak{J}_s^{\nu}(G)$ has been achieved, and if it turns out that

$$\mu_1^{-\nu s}\ \mathfrak{J}_s^{\nu}(G) \leqslant 1,$$

then we will have:

$$p < s.$$

If, on the contrary,

$$\mu_1^{-\nu s}\ \mathfrak{J}_s^{\nu}(G) \leqslant q,$$

with $q > 1$, indicating by h the largest integer such that

$$\binom{s+h}{h} < q,$$

we will have:

$$p \leqslant s + h.$$

Notice that if we dispose of the sequence G_ρ mentioned on page 63, the upper approximation to $\mu_1^{-\nu s}\ \mathfrak{J}_s^{\nu}(G)$ is provided by the numerically known quantity

$$[\mu_1^{(l)}]^{-\nu s}\ \mathfrak{J}_s^{\nu}(G_\rho),$$

where $\mu_1^{(l)}$ is the maximal eigenvalue of the operator $P_l G P_l$, considered on page 33.

The lower approximation to p offers greater difficulties. As a matter of fact, it is possible to construct the sequence $\{\varphi_\nu(G)\}$, setting for instance

$$\varphi_\nu(G) = \mu_1^{\nu - 2\nu^2}\ [\ \mathfrak{J}_1^{\nu^2}\ (G)]^2\,[\ \mathfrak{J}_1^{\nu}(G)]^{-1},$$

or

$$\varphi_\nu(G) = \mu_1{}^{-2\nu}[\ \mathfrak{J}_1^{\nu}(G)]^{\nu+1}[\ \mathfrak{J}_1^{\nu-1}(G)]^{-\nu}. \qquad (1.27.3)$$

Both sequences satisfy the conditions

$$\lim_{\nu \to \infty} \varphi_\nu(G) = p,$$

$$\varphi_\nu(G) < \varphi_{\nu+1}(G),$$

the second one, however, is satisfied only for $\nu > \nu_0$, ν_0 being a *suitable* value of ν, whose actual determination seems very difficult to accomplish.

For instance, for the case of (1.27.3), all that can be said, assuming $G \in \mathfrak{C}^1$, is that

---

[26] See page 608 of [123].

$$\nu_0 > \frac{\mathfrak{I}_1^1(G)}{\mu_1 - \mu_{p+1}} .$$

We shall not consider the case $k > 1$ which, on the other hand, can be reduced to the case $k = 1$, provided it is possible to overcome some additional difficulties. We want, on the contrary, to treat a problem which, apparently far aside from that comprising the computation of the multiplicities of the eigenvalues is actually a particular case of this last problem. *We mean the computation of the Betti numbers of a compact and orientable differentiable manifold.*

In order to state the problem in a precise way, it is convenient to warn that the above computation must be obtained using as data only an oriented atlas and the *passage homeomorphisms* between any two maps of this atlas.

It must be mentioned that, from this standpoint, a solution of the problem has been given by Colautti [124]. Assuming of class C^∞ the r-dimensional, compact and orientable manifold V^r, Colautti considers an arbitrary system $\{\omega_i\}$ of exterior differential forms of degree k (briefly, k-forms), with $0 \leqslant k \leqslant r$, complete with respect to the C^∞ approximation of any C^∞ k-form defined on V^r. Introducing on V^r a Riemannian metric and indicating by $\Delta = d\delta + \delta d$ the Laplace–Beltrami operator for the k-forms relative to such a metric, set:

$$\Delta_1 = \Delta + I \qquad (I \equiv \text{identity operator}).$$

Let $\{v_i^n\}$ be the system obtained by $\{\omega_i\}$, applying the Schmidt orthonormalization procedure to $\{\Delta_1^n \omega_i\}$. If b_k is the Betti number of V^r of the dimension k, one has:

$$b_k = \sum_{i=1}^{\infty} \| v_i^{n_0} \|^2 - \sum_{h,i}^{1,\infty} (\| \Delta_1 v_i^{n_0+h} \|^2 - \| v_i^{n_0+h} \|^2), \tag{1.27.4}$$

where the norm $\| \ \|$ corresponds to the usual scalar product of the $\mathfrak{L}^2$ space of the k-forms on V^r

$$(u, v) = \int_{V^r} u \wedge * v$$

and

$$n_0 > \frac{r}{4}.$$

The depth of the result given by (1.27.4) is illuminated by the fact that it immediately yields one of the most significant theorems of compact manifolds theory: the Poincaré duality theorem, which states that

$$b_k = b_{r-k}. \tag{1.27.5}$$

For, to compute b_{r-k} through (1.27.4), the system $\{*v_i^n\}$ may be employed. But, since the interchange of v_i^n and $*v_i^n$ leaves the right-hand side of (1.27.4) invariant, (1.27.5) is achieved.

The result given by (1.27.4), however, cannot yield bounds on b_k, which, from the numerical standpoint, are much more significant than a convergent algorithm (the quantity b_k to be approximated being an integral!).

Let us consider here a different solution of the problem posed above which not only approximates b_k, but also yields bounds, at least from above, on such a number.

To this end, consider the *kernel double k-form* $H(x, y)$, defined in [125] (*see* page 211) which, in a local coordinate system of V^r, is represented in the following way:

$$H(x, y) = \frac{1}{(k!)^2} h_{s_1 \ldots s_k i_1 \ldots i_k} (x, y) \, dx^{s_1} \cdots dx^{s_k} dy^{i_1} \cdots dy^{i_k}.$$

$H(x, y)$ is a parametrix very simple to construct, to be considered explicitly known as soon as an oriented atlas for V^r and the covariant components of the metric tensor have been assigned. One has

$$\begin{gathered} d_x H(x, y) = \mathcal{O}(|x - y|^{1-r}) \qquad \delta_x H(x, y) = \mathcal{O}(|x - y|^{1-r}) \\ \Delta_x H(x, y) = \mathcal{O}(|x - y|^{1-r}), \end{gathered} \tag{1.27.6}$$

where the distance $|x - y|$ of two (sufficiently close) points of V^r is estimated in an Euclidean way, in an arbitrary system of local coordinates. One has, moreover, for any k-form of class C^∞ on V^r,

$$v(y) = \int (v(x) \wedge *\Delta_x H(x, y) - H(x, y) \wedge *\Delta_x v), \tag{1.27.7}$$

the integration being extended to the oriented manifold V^r. Let S be the $\mathfrak{L}^2$ space of the k-forms on V^r. Consider in S the operator

$$Ku = \int u(x) \wedge *\Delta_x H(x, y)$$

which is compact by (1.27.6). The equation

$$Ku - u = 0 \tag{1.27.8}$$

has, by (1.27.7), all the harmonic k-forms on V^r among its eigenvectors. Now it is proved in [125] that H may be constructed in such a way that the eigenvectors of the operator K are precisely all the harmonic k-forms on V^r.

It follows that the eigenvectors of the problem

$$(K^* - I)(K - I)\,u = 0,$$

where K^* is the adjoint operator to K in S, are none other than the harmonic k-forms on V^r. Set:

$$T = -K^*K + K^* + K.$$

This is a symmetric compact operator, given by an integral kernel in $\mathfrak{L}^2$. A positive number a is easily constructed, in such a way that the spectrum of the operator T is completely contained in the interval $(-a, a)$. We can, of course, assume $a > 1$.

Now let $p(\lambda)$ be a polynomial in λ, with real coefficients, satisfying all the following conditions:

$$0 < p(-a) < 1, \qquad 0 < p(a) < 1,$$

$$p(0) = 0, \qquad p'(0) = 0, \qquad p(1) = 1, \qquad p'(1) = 0.$$

$$p'(x) < 0 \quad \text{for} \quad -a < x < 0,$$

$$p'(x) > 0 \quad \text{for} \quad 0 < x < 1,$$

$$p'(x) < 0 \quad \text{for} \quad 1 < x < a.$$

Set

$$G = p(T).$$

The operator G is compact and symmetric. G, in addition, is positive, its spectrum being wholly contained in the interval (0, 1). If there exist harmonic k-forms on V^r which do not vanish identically, these are none other than the eigenvectors of the operator G, corresponding to the largest eigenvalue $\mu_1 = 1$.

Since the integral representation of G is explicitly known, its orthogonal invariants $\mathfrak{I}_1^\nu(G)$ may be computed by means of quadratures on V^r. On the other hand, by a fundamental theorem of Hodge, b_k is equal to the maximum number of linearly independent harmonic k-forms existing on V^r. The above-mentioned computation procedures for p may thus be applied. In particular one has

$$b_k = \lim_{\nu \to \infty} \mathfrak{I}_1^\nu(G),$$

and it turns out that

$$b_k < \mathfrak{I}_1^\nu(G).$$

It is to be noticed that this bound holds for any parametrix H, i.e. even if (1.27.8) has some eigenvectors different from the harmonic k-forms.

28. Approximations and Estimates for Eigenvectors

The methods and the applications, summarized in the previous Sections, refer to the eigenvalues. The problem of the approximation of the corresponding eigenvectors has been investigated by several authors from different points of view (*see* [189] for references). We wish to relate here some general results which have been obtained quite recently [189], in connection with positive compact operators, concerning eigenvectors approximation.

It is natural to assume as 'approximate eigenvector' of the positive compact operator G, introduced in Section 2, corresponding to the eigenvalue μ_k ($k = 1, 2, \ldots$), the eigenvector given by the Rayleigh–Ritz method and corresponding to $\mu_k^{(\nu)}$, i.e. the vector

$$u^{(\nu)} = \gamma_1 w_1 + \ldots + \gamma_\nu w_\nu$$

where $(\gamma_1, \ldots, \gamma_\nu)$ is a non-trivial solution of the homogeneous system:

$$\sum_{j=1}^{\nu} (\mathrm{G}w_j, w_k)\,\gamma_j - \mu_k^{(\nu)}(w_j, w_h)\,\gamma_j = 0 \qquad (h = 1, \ldots, \nu).$$

However, even if we suppose that $\| u^{(\nu)} \| = 1$, the sequence $\{u^{(\nu)}\}$ is, in general, non-converging. That is very easily seen. However, in [189], as a consequence of general results, it is proved that:

1. XIX. *The sequence* $\{u^{(\nu)}\}$ *is compact*[27] *and every compactness vector for* $\{u^{(\nu)}\}$ [28] *is an eigenvector for* G, *corresponding to the eigenvalue* μ_k.

Since $\{u^{(\nu)}\}$ is non-converging, the problem of the *estimation of the error* has no meaning. However, since all the compactness vectors of $\{u^{(\nu)}\}$ belong to the subspaces U_k of S, formed by the solutions of the equation $\mathrm{G}u - \mu_k u = 0$, a very important problem is the one consisting in *estimating the distance of* $u^{(\nu)}$ *from* U_k.

If P is the (orthogonal) projector of S onto U_k and we set $\mathrm{Q} = \mathrm{I} - \mathrm{P}$, the distance of $u^{(\nu)}$ from U_k is the norm $\| \mathrm{Q}u^{(\nu)} \|$ of the vector $\mathrm{Q}u^{(\nu)}$.

Suppose that μ_k has multiplicity p and $\mu_{k-1} > \mu_k, \mu_k > \mu_{k+p}$. In [189] the following inequality, which estimates the distance of any vector v of S from U_k, is proved:

$$\| \mathrm{Q}v \| \leqslant \frac{[\| \mathrm{G}v \|^2 - 2\mu_k(\mathrm{G}v, v) + \mu_k^2 \| v \|^2]^{1/2}}{\min\{\mu_{k-1} - \mu_k, \mu_k - \mu_{k+p}\}}. \qquad (1.28.1)$$

In order to apply (1.28.1) in practice, we need to know:

(a) the multiplicity p of μ_k;

(b) a lower bound μ'_{k-1} for μ_{k-1} and an upper bound μ''_k for μ_k such that $\mu'_{k-1} > \mu''_k$;

(c) a lower bound μ'_k for μ_k and an upper bound μ''_{k+p} for μ_{k+p}, such that $\mu'_k > \mu''_{k+p}$.

Then (1.28.1) gives:

$$\| \mathrm{Q}v \| \leqslant \frac{[\| \mathrm{G}v \|^2 - 2\mu'_k(\mathrm{G}v, v) + \mu''_k \| v \|^2]^{1/2}}{\min\{\mu'_{k-1} - \mu''_k, \mu'_k - \mu''_{k+p}\}}. \qquad (1.28.2)$$

Procedures for obtaining the required lower and upper bounds and p have been described in the previous Sections.

In [189] the problem of the approximation of the eigenvectors has been considered also from a 'global' point of view. To this end the sequence $\{\mathrm{U}^{(\nu)}\}$ of subspaces of S is

(27) By stating that $\{u^{(\nu)}\}$ is compact we mean that any subsequence of $\{u^{(\nu)}\}$ contains a converging subsequence.

(28) A *compactness vector* for $\{u^{(\nu)}\}$ is the limit of any converging subsequence of $\{u^{(\nu)}\}$.

said to converge (to converge uniformly) to the subspace U if the projector Π_ν of S onto $U^{(\nu)}$ converges strongly (converges uniformly) to the projector Π of S onto U.

Consider the sequence $\{G_\nu\}$ of positive compact operators and suppose that $\{G_\nu\}$ converges uniformly to the positive compact operator G. Consider the positive eigenvalues $\mu_k, \ldots, \mu_{k+p-1}$ $(p \geqslant 1)$ of G and let U be the space spanned by the corresponding eigenvectors. Let $\mu_k^{(\nu)}, \ldots, \mu_{k+p-1}^{(\nu)}$ be eigenvalues of G_ν and $U^{(\nu)}$ the space spanned by the corresponding eigenvectors. The following theorem answers the question of the convergence of $\{U^{(\nu)}\}$ towards U, by showing that the projector Π of S onto U is a Lipschitz–continuous function of G in the Banach space of bounded operators in S (*see* [189]).

1.XX. *Let* G *and* G′ *be two positive compact operators. Denote by* $\{\mu_k\}, \{u_k\}$ *and* $\{\mu'_k\}, \{u'_k\}$ *the eigenvalues and the corresponding eigenvectors (orthonormalized) of* G *and* G′, *respectively. Suppose that* $\mu_{k-1} > \mu_k \geqslant \ldots \geqslant \mu_{k+p-1} > \mu_{k+p}$ $(\mu_{k-1} = +\infty$ *if* $k = 1)$ *and, moreover, that*

$$\| G - G' \| \leqslant \frac{1}{4} \min \{\mu_{k-1} - \mu_k, \mu_{k+p-1} - \mu_{k+p}\}.$$

If U *is the space spanned by* $u_k, \ldots, u_{k+p-1}$, U′ *the space spanned by* $u'_k, \ldots, u'_{k+p-1}$, Π *and* Π' *the projectors onto* U *and* U′, *respectively, then*

$$\| \Pi' - \Pi \| \leqslant \frac{2(2p)^{1/2}(\| G \| + \mu_k)}{\min \{(\mu_{k-1} - \mu_k)^2, (\mu_{k+p-1} - \mu_{k+p})^2\}} \| G' - G \|.$$

29. An Eigenvalue Problem Proposed by Ostrowski

Let us consider, in the space $\mathfrak{L}^2[0, 1]$, the operator

$$Gu = \int_0^1 K(x, y)\, u(y)\, dy \tag{1.29.1}$$

where

$$K(x, y) \begin{cases} = 2 \log \dfrac{x(1-y)}{x-y} & \text{for } x > y \\ \\ = 2 \log \dfrac{y(1-x)}{y-x} & \text{for } y > x. \end{cases} \tag{1.29.2}$$

The operator G is a positive compact operator belonging to $\mathfrak{C}^2$.

It has been shown that the greatest eigenvalue μ_1 of G coincides, for $\alpha = 2$, with the smallest constant c_α, such that the following inequality holds:

$$\int_0^1 \int_0^1 \left| \frac{f(x)-f(y)}{x-y} \right|^\alpha dx\,dy \leqslant c_\alpha \int_0^1 |f'(x)|^\alpha dx \qquad (\alpha \geqslant 1) \tag{1.29.3}$$

where $f(x)$ is any absolutely continuous function in [0, 1], such that $f'(x) \in \mathfrak{L}^\alpha\,[0, 1]$ (*see* [190]).

Ostrowski, who has proved that: (a) c_α is a non-increasing function of α; (b) $c_1 = \log 4$, $c_\alpha \geqslant 1$; (c) $c_\alpha \leqslant c_2^{\alpha-1} c_1^{2-\alpha}$ for $1 < \alpha < 2$; (d) $c_\alpha \geqslant c_2^{\alpha-1} c_1^{2-\alpha}$ for $\alpha > 2$, has posed the problem of the rigorous computation of c_2, i.e. μ_1. This problem has been solved in [191].

Assuming $w_i(x) = x^{i-1}$ $(i = 1, 2, \ldots)$ let us denote by $\mu_1^{(\nu)} \geqslant \ldots \geqslant \mu_\nu^{(\nu)}$ the roots of the determinant equation (1.9.6) where, now (,) denotes the standard scalar product in $\mathfrak{L}^2\,[0, 1]$.

Let

$$M_1^{(\nu)} \geqslant \ldots \geqslant M_\nu^{(\nu)}$$

be the roots of the determinant equation:

$$\det\, \{(G^2 w_i, w_h) - M\,(w_i, w_h)\} = 0 \qquad (i, h = 1, \ldots, \nu).$$

We have, for the eigenvalue μ_k of G,

$$\mu_k^{(\nu)} \leqslant (M_k^{(\nu)})^{1/2} \leqslant \mu_k \qquad (k = 1, \ldots, \nu).$$

Upper bounds for μ_k can be obtained through (1.9.7) which we have used in the case $s = 1, n = 3$.

For the operator G we have:

$$\mathfrak{I}_1^3(G) = \frac{70}{9}\pi^2 - \frac{194}{3} - 8\zeta(3)$$

where ζ is the Riemann zeta-function.

A refinement of (1.9.7), in the case $s = 1$, has been obtained in [191] which permits an improvement in the upper bounds.

A further improvement of the upper bound for μ_1 is given by the following inequality:

$$\mu_1 \leqslant \frac{M_1^{(\nu)} - \sigma_2^{(\nu)} \mu_1^{(\nu)}}{\mu_1^{(\nu)} - \sigma_2^{(\nu)}}$$

where $\sigma_2^{(\nu)}$ is given by (1.9.7) or by analogous, more refined, formulae and ν is such that $\mu_1^{(\nu)} > \sigma_2^{(\nu)}$ [192], [191].

For the eigenvalue μ_1 the following estimates were obtained:

$$1.202931525711 < \mu_1 < 1.202931525733.$$

30. Miscellaneous Results Concerning Eigenvalue Theory

We shall rapidly review in this Section several results concerning eigenvalue theory, which, unlike those examined in the former Sections, have an isolated character and are not inserted in the framework of a unified theory.

Several results arose from the attempts at tackling the lower approximation of the eigenvalues before the elaboration of the orthogonal invariants method.

De Vito [126] frames into an abstract setting several computation procedures for the eigenvalues known at that time. He gives an interesting treatment of the *method of Temple*, who made the first attempt to compute lower bounds to eigenvalues in some generality. In the scheme and notations of Section 2, the Temple method is summarized by the following theorem:

1.XXI. *Let σ be a real number such that $\lambda_k < \sigma \leqslant \lambda_{k+1}$. For any $v \in \mathrm{V}$, such that* $(v, \mathrm{L}v - \sigma v) \neq 0$, *the functional*

$$\mathrm{F}_\sigma(v) = \frac{(\mathrm{L}v, \mathrm{L}v - \sigma v)}{(v, \mathrm{L}v - \sigma v)}$$

is well defined. Let D_σ be the set of vectors of V *such that* $(v, \mathrm{L}v - \sigma v) < 0$. *The maximum in D_σ of the functional $\mathrm{F}_\sigma(v)$ is then the eigenvalue λ_k.*

It is seen that this theorem yields a method for obtaining lower bounds to λ_k *as soon as a number σ 'separating' λ_k and λ_{k+1} is known;*(29)

Theorem 1.XXI is related in [126] to a general theorem dealing with the separation of the eigenvalues, which, proved by Wielandt [127] for Hermitian matrices and by Bückner [128] for Fredholm integral operators, is demonstrated in [129] for the general operator L of Section 2 of the present treatment.

Set $\mathrm{V}_0 = \mathrm{V}$, and, by induction, $\mathrm{V}_k = \mathrm{G}(\mathrm{V}_{k-1})$.

The general theorem is:

1.XXII. *Let v be a non-zero vector of V_{n-1} and $g(x) = \sum_{k=0}^{n} a_k x^k$ be a polynomial having real coefficients such that*

$$a_n(\mathrm{L}^n v, v) + a_{n-1}(\mathrm{L}^{n-1} v, v) + \ldots + a_1(\mathrm{L}v, v) + a_0\| v \|^2 = 0.$$

Denoting by $\mathrm{R}_+(g)$ the set of the points on the real axis such that $g(x) \geqslant 0$ and by $\mathrm{R}_- g$ the one such that $g(x) \leqslant 0$, both sets contain eigenvalues of Problem (1.2.1).

For the validity of Theorem 1.XXII it is enough to suppose that $v \in \mathrm{V}_{p_n - 1}$ where p_n is the smallest integer such that $p_n \geqslant n/2$.

This theorem, whose interest is apparent, did not attract much attention. Recently for instance L. Fox, Henrici and Moler [130] have obtained interesting numerical results

---

(29) Theorem 1.XXI is included also in the recent monograph [33], where however the vector v is not subject to the condition $(v, \mathrm{L}v - \sigma v) < 0$, in order to have $\mathrm{F}_\sigma(v) < \lambda_k$. As a matter of fact the proof contained in pages 103 and 104 of [33] is incorrect, since the two equations $\tau + \varphi(\tau) = \rho$ and $(\tau - \rho)^2 = [\varphi(\tau)]^2$ of page 103 are assumed to be equivalent.

in some eigenvalue problems by means of the following lemma:

If $v \neq 0$, $v \in V$, *and if one sets*

$$\rho = \frac{(v, Lv)}{\| v \|^2}, \qquad \sigma = \frac{(Lv, Lv)}{\| v \|},$$

there exists some λ_k *such that*

$$\rho - \sqrt{\sigma^2 - \rho^2} \leqslant \lambda_k \leqslant \rho + \sqrt{\sigma^2 - \rho^2}.$$

Such a lemma, as a matter of fact, is a very particular case of theorem 1.XXII, since the polynomial

$$g(x) = x^2 - 2\rho x + 2\rho^2 - \sigma^2$$

satisfies the conditions of the theorem and $R_-(g)$ is, in the present case, the interval

$$[\rho - \sqrt{\sigma^2 - \rho^2}, \rho + \sqrt{\sigma^2 - \rho^2}].$$

The Note [131] deals with the lower approximation of the lowest eigenvalue for problems connected with second order elliptic equations. More precisely, the eigenvalue problem under consideration is the following:

$$\sum_{hk}^{1,r} \frac{\partial}{\partial x_h} \left(a_{hk} \frac{\partial u}{\partial x_k} \right) - bu + \lambda u = 0 \qquad \text{in A}$$

with mixed boundary conditions

$$u = 0 \qquad \text{on } \partial_1 \text{A}, \qquad \sum_{hk}^{1,r} a_{hk} \nu_h \frac{\partial u}{\partial x_k} - pu = 0 \qquad \text{on } \partial_2 \text{A},$$

A being the usual bounded properly regular domain, $\sum_{h,k}^{1,r} a_{hk}(x)\, \xi_h\, \xi_k$ being positive definite ($a_{hk} = a_{kh}$), $b(x) > 0$ in $\bar{\text{A}}$ and $p \geqslant 0$ on ∂_1A. Furthermore

$$\partial \text{A} = \overline{\partial_1 \text{A}} \cup \overline{\partial_2 \text{A}}, \qquad \partial_1 \text{A} \cap \partial_2 \text{A} = \phi.$$

The functions a_{hk}, b and p must in addition satisfy suitable smoothness assumptions. Denoting by $\{\alpha_{hk}\}$ the inverse of the $r \times r$ matrix $\{a_{hk}\}$ and by φ an r-vector, consider the functional

$$\text{P}(\varphi) = \inf_{x \in \text{A}} \left[b(x) + \sum_{h=1}^{r} \frac{\partial \varphi_h}{\partial x_h} - \sum_{hk}^{1,r} \alpha_{hk} \varphi_h \varphi_k \right].$$

Let Φ be the class of the r-vectors belonging to $C^1(\bar{\text{A}}) \cap C^0(\text{A})$ such that $\varphi\nu + p \geqslant 0$ on ∂_2A.

It has been proved by several authors (Boggio, Picard, Barta, Protter, Hersch, Picone) that

$$P(\varphi) \leqslant \lambda_1 \quad \text{for } \varphi \in \Phi.$$

It is shown in [131] that

$$\sup_{\varphi \in \Phi} P(\varphi) = \lambda_1.$$

The result is then extended to the problem in which the parameter λ occurs also in the boundary condition.

It is worth remarking that the lower approximation of λ_1 through $P(\varphi)$ is actually possible only in order to obtain rough lower bounds, the method being ineffective in producing sharper results.

Finally, let us mention a recent result obtained by Sneider [132]. Consider the Fredholm integral equation

$$\int_0^1 K(x, y)\, u(y)\, dy = \mu u(x) \tag{1.29.1}$$

with a (not necessarily Hermitian) kernel such that

$$0 < \int_0^1 \int_0^1 |K(x, y)|^2\, dx\, dy < +\infty. \tag{1.29.2}$$

Sneider proves that, denoting by $\sigma(K)$ the area of the circular domain of minimal area containing *all* eigenvalues of (1.29.1) and considering the functional

$$I(K) = \frac{\sigma(K)}{\int_0^1 \int_0^1 |K(x, y)|^2\, dx\, dy}.$$

within the class of the functions $K(x, y)$ satisfying (1.29.2), one has

$$\max I(K) = \frac{\pi}{2}.$$

This author proves, moreover, that a maximizing function is, for instance, the following one:

$$K(x, y) = 2\,(\sin \pi x \sin \pi y - \sin 2\pi x \sin 2\pi y).$$

Sneider's theorem leads to the following (not improvable) bound on the radius r of the minimal circle containing all eigenvalues of (1.29.1):

$$r^2 \leqslant \frac{1}{2} \int_0^1 \int_0^1 |K(x, y)|^2\, dx\, dy.$$

Part II: 'A Priori' Estimates for Solutions of Differential Equations

1. Global Estimates Connected with a Class of Linear Transformations

The investigations we shall report in this Section began, when the writer was working at the Istituto Nazionale per le Applicazioni del Calcolo (Rome), with the intention of obtaining majorization formulae in order to estimate the approximation error of the least square methods, often employed in that Institute. Already in [133] the following estimate was proved

$$\int_A u^2\, dx \leqslant K \int_{\partial A} u^2\, d\sigma$$

valid for an harmonic function in a bounded domain A, with a regular boundary. However the computation of K was achieved through empirical procedures. The question has been subsequently reinvestigated in a more general setting in the memoir [134], whose results we now come to develop.

Let S and S′ be two linear Banach spaces over the same numerical field N and let

$$u' = T(u) \tag{2.1.1}$$

be a linear transformation from S into a linear manifold T(S) of S′.

Assume the existence and the continuity of the inverse transformation of (2.1.1):

$$u = T^{-1}(u') \tag{2.1.2}$$

which is of course linear.

We want to construct, under suitable assumptions on the given transformation and its inverse, a majorization method for the norm of (2.1.2), i.e. a procedure for obtaining upper bounds to the supremum $\mu(T^{-1})$ of the functional

$$\frac{\| T^{-1}(u')\|}{\| u' \|} \qquad [u' \subset T(S)].$$

The class of transformations we shall consider includes one relative to a set of transformations connected with boundary problems for linear differential equations, which in turn include those relative to the classical boundary problems for second order elliptic or parabolic equations.

Incidentally, a further contribution will be made to a result of Caccioppoli, concerning a global estimate for the solutions of the linear differential equations of elliptic type [135].

Let $B_1, B_2, \ldots, B_n, G_1, G_2, \ldots, G_m$ be $n + m$ Borel sets belonging to the Euclidean spaces $S_{k_1}, S_{k_2}, \ldots, S_{k_n}, S_{h_1}, S_{h_2}, \ldots, S_{h_m}$ respectively.

Let S_B be a linear manifold, over the real field N, formed by the vectors $u \equiv (u_1, u_2, \ldots, u_n)$, with n real components, the i-th component being defined in B_i and S_G a linear manifold composed of vectors $U \equiv (U_1, U_2, \ldots, U_m)$, U_j being real and defined in G_j.

Let there be defined, in the linear manifold $S_B^{(0)}$ contained in S_B, the linear transformations

$$w = E(u), \qquad w = E^*(u)$$

having range

$$E(S_B^{(0)}), \qquad E^*(S_B^{(0)})$$

contained in S_B. Further, let there be defined in $S_B^{(0)}$ the three linear transformations

$$U = L(u), \qquad U = L^*(u), \qquad U = M(u)$$

with ranges

$$L(S_B^{(0)}), \qquad L^*(S_B^{(0)}), \qquad M(S_B^{(0)})$$

contained in S_G.

We shall assume that in B_i and G_j are defined the two non-negative completely additive set functions $\beta_i(E_i)$ and $\gamma_j(I_j)$, for all Borel sets E_i and I_j, contained in B_i and G_j, respectively.

Let the vectors u of S_B and U of S_G be such that the Lebesgue–Stieltjes integrals

$$\int_{B_i} u_i^2 \, d\beta_i, \qquad \int_{G_j} U_j^2 \, d\gamma_j,$$

have finite values.

Denoting by u and v two vectors of S_B, let us set

$$(u, v)_B = \sum_{i=1}^{n} \int_B u_i v_i \, d\beta_i$$

and, denoting by U and V two vectors of S_G, set

$$(U, V)_G = \sum_{j=1}^{m} \int_{G_j} U_j V_j \, d\gamma_j.$$

S_B and S_G are therefore Hilbert spaces.

Let us state the following first hypothesis.

(1) *For any pair of vectors u and v in $S_B^{(0)}$, the following identity holds:*

$$(u, E^*(v))_B - (v, E(u))_B = (M(u), L^*(v))_G - (M(v), L(u))_G.$$

Let S′ be the Hilbert space $S_B \times S_G$, whose elements will be denoted by $u' \equiv (u, U)$. Consider the two linear transformations from $S \equiv S_B^{(0)}$ in S′:

$$u' = T(u) \equiv (E(u), L(u)); \qquad u' = T^*(u) \equiv (E^*(u), L^*(u))$$

and let us state the following second hypothesis:

(2) *Let there exist the inverse transformations of* $T(u)$ *and* $T^*(u)$:

$$u = T^{-1}(u'), \qquad u = T^{*-1}(u').$$

Denote by ω the null element in S_B and by Ω the null element in S_G and set:

$$u_1' \equiv (u, \Omega), \qquad u_2' \equiv (\omega, U)$$

so that

$$u' = u_1' + u_2'.$$

Let us further set:

$$E^{-1}(u) = T^{-1}(u_1') \quad , \qquad L^{-1}(U) = T^{-1}(u_2')$$
$$E^{*-1}(u) = T^{*-1}(u_1'), \qquad L^{*-1}(U) = T^{*-1}(u_2')$$

and hence

$$T^{-1}(u) = T^{-1}(u_1') \quad , \qquad L^{-1}(U) = T^{-1}(u_2')$$
$$E^{*-1}(u) = T^{*-1}(u_1'), \qquad L^{*-1}(U) = T^{*-1}(u_2').$$

Let us consider the following third hypothesis:

(3) *Let there exist a transformation from* S_B *into a subset of it, containing* $E(S_B^{(0)})$,

$$w = \varphi(u)$$

and a transformation from S_G *into a subset of it, containing* $L(S_B^{(0)})$

$$W = \Phi(U)$$

such that one has:

$$0 \leqslant w_i \leqslant |u_i|, \qquad 0 \leqslant W_j \leqslant |U_j|$$

$$(i = 1, 2, \ldots, n; j = 1, 2, \ldots, m)$$

Furthermore, setting

$$(E_1^{-1}(u), E_2^{-1}(u), \ldots, E_n^{-1}(u)) \equiv E^{-1}(u)$$
$$(L_1^{-1}(U), L_2^{-1}(U), \ldots, L_n^{-1}(U)) \equiv L^{-1}(U)$$

let us in addition have

$$E_i^{-1}(\varphi(u)) \geqslant E_i^{-1}(u)$$
$$L_i^{-1}(\Phi(U)) \geqslant L_i^{-1}(U).$$

We can, now, state the theorem:

2.I *Let the functional*

$$I(u) = \frac{\| u \|^2}{\| E(u)\|^2 + \| L(u)\|^2}$$

have a maximum in $S_B^{(0)}$ *and let there exist a maximizing vector* u_0 *such that* $T(u_0)$ *has non-negative components. Choosing a vector* w *of* $S_B^{(0)}$ *satisfying the conditions*

$$L^*(w) = \Omega, \qquad E_i E^*(w) > 0, \qquad L_j E^*(w) > 0$$

and denoting by $F(w)$ *the maximum of the numbers*

$$\sup_{B_i} \frac{w_i}{E_i E^*(w)}, \qquad \inf_{G_j} \frac{M_j(w)}{L_j E^*(w)},$$

for any $u \in S_B^{(0)}$, *one has*:

$$\| u \|^2 \leqslant F(w) [\| E(u)\|^2 + \| L(u)\|^2].$$

Among the several applications of this theorem, let us consider the following one:

2.II. *Let* D *be a bounded closed domain of* S_r *and let*

$$E(u) = \sum_{hk}^{1,r} a_{hk} \frac{\partial^2 u}{\partial x_h \partial x_k} + \sum_{1}^{r}{}_k\, b_k \frac{\partial u}{\partial x_k} + cu$$

be a second order positive elliptic differential operator. Let the boundary ∂D *of* D *be of class* 2 *at any point and let the coefficients* a_{hk} *be of class* 2 *in* D *with Hölder continuous second derivatives and let* b_k*'s be of class* 1 *with Hölder continuous first derivatives. Let* c *be Hölder continuous and non-positive in* D. *Set*:

$$E^*(u) = \sum_{hk}^{1,r} \frac{\partial^2}{\partial x_h \partial x_k} (a_{hk} u) - \sum_{k}^{1,r} \frac{\partial b_k u}{\partial x_k} + cu$$

and let w *be an arbitrary function of class* 2 *in* D, *such that* $E^*(w)$ *is of class* 2 *in* D, *satisfying, in addition, the following conditions*:

$$w = 0 \quad \text{on } \partial D, \qquad E^*(w) < 0 \quad \text{on } \partial D, \tag{2.1.3}$$

$$EE^*(w) > 0 \quad \text{in } D. \tag{2.1.4}$$

Denoting by $F(w)$ *the maximum of the two numbers*:

$$\max_{D} \frac{w}{EE^*(w)}, \qquad \max_{\partial D} \frac{\partial w/\partial \nu}{E^*(w)} \qquad (\nu \text{ outer conormal to } \partial D)$$

then for any function of class 1 *in* D, *with second derivatives continuous in* $D - \partial D$ *and square integrable in* D, *the following majorization holds*:

$$\int_D u^2\,d\tau \leqslant F(w)\left\{\int_D |E(u)|^2\,d\tau + \int_{\partial D} u^2\,d\sigma\right\}. \tag{2.1.5}$$

If w*, besides* (2.1.3), *satisfies the condition*

$$EE^*(w) = 0 \qquad \textit{in } D, \tag{2.1.4'}$$

then setting:

$$F(w) = \max_{\partial D} \frac{\partial w/\partial \nu}{E^*(w)},$$

for any solution u *of class* 1 *in* D *of the equation* $E(u) = 0$, *the following majorization holds*:

$$\int_D u^2\,d\tau \leqslant F(w)\int_{\partial D} u^2\,d\sigma. \tag{2.1.5'}$$

Finally, if w *stands for a function such that*

$$w = 0 \textit{ on } \partial D, \qquad E^*(w) = 0 \textit{ on } \partial D, \qquad EE^*(w) > 0 \textit{ in } D,$$

then setting

$$F(w) = \max_{D} \frac{w}{EE^*(w)},$$

the following estimate holds for any function u *continuous in* D, *vanishing on* ∂D, *of class* 1 *in* D, *having continuous second derivatives in* $D - \partial D$ *and square integrable in* D:

$$\int_D u^2\,d\tau \leqslant F(w)\int_D |E(u)|^2\,d\tau. \tag{2.1.5''}$$

In the paper quoted above Caccioppoli proved (2.1.5′) and (2.1.5″), taking, as $F(w)$, $4\max_D v^2$ in (2.1.5″), v being the solution of $E^*(v) = 1$ vanishing on ∂D and taking in (2.1.5′) $F(w)$ the maximum of $\partial v/\partial \nu$ on ∂D.

Theorem 2.II represents a further contribution to Caccioppoli's result, from the quantitative standpoint, since, in order to achieve the majorization, it is not necessary to have knowledge of v, which is, in general, not easy to obtain but only of a function such as w in the theory which, on the contrary, may be constructed in a simple way.

If, for instance, $E(u)$ is the two-variables Laplace operator $\Delta_2 u = \dfrac{\partial u^2}{\partial x_1^2} + \dfrac{\partial^2 u}{\partial x_2^2}$ and D the elliptic domain

$$\frac{x_1^2}{a^2}+\frac{x_2^2}{b^2}\leqslant 1 \qquad (a\geqslant b>0),$$

to achieve (2.1.5) one may assume

$$w=\left(\frac{x_1^2}{a^2}+\frac{x_2^2}{b^2}\right)-8\left(\frac{x_1^2}{a^2}+\frac{x_2^2}{b^2}\right)+7,$$

so that $F(w)$ will be the maximum of the two numbers:

$$\frac{7a^4b^4}{24(a^4+b^4)+16a^2b^2},\qquad \frac{3a^2b}{a^2+3b^2}.$$

If, on the contrary, D is the rectangular domain

$$-a\leqslant x_1\leqslant a,\qquad -b\leqslant x_2\leqslant b,$$

always with $E(u)\equiv\Delta_2 u$,[1] assume:

$$w=(x^2-a^2)(y^2-b^2)$$

and, hence, for $F(w)$ the maximum among

$$a,\qquad b,\qquad \frac{a^2b^2}{8}.$$

Results analogous to those of Theorem 2.II may be achieved for equations of parabolic type.

2.III. *Let* D *be the domain of* S_r *considered in Theorem* 2.II *and let* T *be a cylindrical domain of the space* $S_{r+1}\equiv(x_1,x_2,\ldots,x_r,t)$, *defined by*

$$x=(x_1,x_2,\ldots,x_r)\subset D,\qquad 0\leqslant t\leqslant t_0.$$

We shall denote by $\partial_1 T$, $\partial_2 T$, $\partial_3 T$ *those parts of the boundary of* T *defined, respectively, by the conditions*

$$t=0,\qquad x\subset\partial D,\qquad t=t_0.$$

Considering the r-variables elliptic differential operator $E(u)$ *introduced in Theorem* 2.II, *let us set*:

$$P(u)=E(u)-\frac{\partial u}{\partial t},\qquad P^*(u)=E^*(u)+\frac{\partial u}{\partial t}$$

and let us denote by w *an arbitrary function of class* 2 *in* T, *such that* $P^*(u)$ *is of class* 2 *in* T, *satisfying in addition the conditions*:

$$w=0 \qquad \text{on } \partial_2T\cup\partial_3T,$$

$$PP^*(w)>0 \qquad \text{in } T,$$

$$P^*(w)<0 \qquad \text{on } \partial_1T\cup\partial_2T.$$

---

[1] Theorem 2.II holds also for a rectangular domain, if $E(u)$ coincides with the Laplace operator $\Delta_2 u$.

Denoting by F(w) *the maximum among the numbers*

$$\max_{\mathrm{T}} \frac{w}{\mathrm{PP}^*(w)}, \qquad \max_{\partial_1 \mathrm{T}} \frac{w}{\mathrm{P}^*(w)}, \qquad \max_{\partial_2 \mathrm{T}} \frac{\partial w/\partial \nu}{\mathrm{E}^*(w)}$$

(ν *being the outward normal to the cross-section* D *of* T), *the following majorization formula holds for any function of class* 1 *in* T, *with second derivatives* (*with respect to* $x_1, x_2, \ldots, x_r$) *continuous in* $\mathrm{T} - \partial_2 \mathrm{T}$ *and square integrable in* T:

$$\int_0^{t_0} \mathrm{d}t \int_{\mathrm{D}} u^2 \,\mathrm{d}\tau \leqslant \mathrm{F}(w) \left\{ \int_0^{t_0} \mathrm{d}t \int_{\mathrm{D}} \left[\mathrm{E}(u) - \frac{\partial u}{\partial t} \right]^2 \mathrm{d}\tau + \int_{\partial_1 \mathrm{T}} u^2 \,\mathrm{d}\tau + \int_{\partial_2 \mathrm{T}} u^2 \,\mathrm{d}\sigma \right\}.$$

If, for instance, we take $r = 1$, D coinciding with the interval $0 \leqslant x \leqslant 1$ and $t_0 = 1$, the theorem just stated leads to the following majorization:

$$\int_0^1 \int_0^1 u^2 \,\mathrm{d}x \,\mathrm{d}y \leqslant \frac{1}{2} \left[\int_0^1 [u(x, 0)]^2 \,\mathrm{d}x + \int_0^1 [u(0, t)]^2 \,\mathrm{d}t \right.$$

$$\left. + \int_0^1 [u(t, 1)]^2 \,\mathrm{d}t + \int_0^1 \int_0^1 \left[\frac{\partial^2 u}{\partial x^2} - \frac{\partial u}{\partial t} \right]^2 \mathrm{d}x \,\mathrm{d}t \right].$$

In [134] a more sophisticated theorem than 2.I is also considered, in which the hypothesis of the existence of the maximum of I(u) is replaced by another less restrictive one.

The results of [134] have been pursued and extended in [136]. Here also second order partial differential systems are considered, which are either elliptic or parabolic. Among the several estimates contained in [136], let us mention here that concerning the elasticity operator $\Delta_2 u + \sigma$ grad div u ($\sigma > -1$) in the space X^r. For this operator the following inequality is proved, valid for any u vanishing on ∂A (A bounded)

$$\int_{\mathrm{A}} |u|^2 \,\mathrm{d}x \leqslant \left(\frac{r\delta^2}{r + \delta} \right)^2 \int_{\mathrm{A}} |\Delta_2 u + \sigma \operatorname{grad} \operatorname{div} u|^2 \,\mathrm{d}x,$$

δ being the radius of any circular domain containing A. It follows easily from such an estimate that, if λ_1 is the lowest eigenvalue of the problem

$$\Delta_2 v + \sigma \operatorname{grad} \operatorname{div} v + \lambda v = 0 \quad \text{in A},$$

$$v = 0 \quad \text{on } \partial\mathrm{A},$$

one has:

$$\lambda_1 \geqslant \frac{r + \sigma}{r\delta^2}.$$

The results achieved in [134] and [136] gave rise to a series of investigations by the 'Maryland School' and, in particular, by Weinberger, Payne, Diaz and Bramble, who extended and refined in various directions the writer's results. For a complete list of references on the subject, see the monograph [137] by Davis and Rabinowitz.

The procedures of [134] and [136] have been applied by Colautti [138] to the case of second order ordinary differential equations and systems, with general boundary conditions.

(For reference to Colautti's work, *see* [139] p. 180.)

In [140] applications may be found to the exterior problems for three-variables harmonic functions.

2. The General Existence Principle in Banach Spaces and the Dual Estimation Formulae

The application of linear functional analysis to the classical problems of differential equations theory must be traced back to Renato Caccioppoli [141] and Herman Weyl [142]. The first author made use of the Hahn–Banach theorem, in order to give a clever proof of the classical existence theorems on a Riemann surface; the second one employed the 'projection theorem' in the Hilbert spaces to obtain the existence proof of the solution of the Dirichlet problem for a second order elliptic equation.

In this framework a particular success has been achieved by a lemma of Lax and Milgram [143], which extends the classical theorem of F. Riesz, on the representation of the continuous linear functionals, on a Hilbert space.

In all these functional analysis theorems one can find, in a more or less explicit way, the fact that, in achieving the existence of the solution, an 'a priori estimate' for the solutions of the linear equation under investigation must be used.

Almost at the same time as Lax and Milgram obtained their result, the writer formulated a general existence principle in Banach spaces, communicated at the International Congress on Partial Differential Equations held in Trieste in 1954, which can be used to prove that the existence of the solution of a linear equation and the existence of an 'a priori' estimate are perfectly equivalent statements. Such a principle, which contains as particular cases all the existence theorems employed by the authors quoted above, can be used in a double way: either in order to deduce existence theorems from the 'a priori' estimates or, conversely, to obtain estimates from these theorems.

Let us give a short exposition of such a principle.

Let V be a complex vector space and let M_1 and M_2 be two linear transformations of domain V and ranges contained in the two Banach spaces B_1 and B_2, respectively.

Let φ be an element of the space B_1^*, the topological dual of B_1. Let us consider the equation

$$\langle \varphi, M_1 v \rangle = \langle \psi, M_2 v \rangle \qquad \text{for any } v \in V, \tag{2.2.1}$$

where the 'unknown' ψ is an element of B_2^* (topological dual of B_2).

The above-mentioned existence principle comprises the following theorem:

2.IV. *A necessary and sufficient condition for the existence of a solution* ψ *of* (2.2.1), *for any* φ, *is that there exists a constant* $K > 0$, *such that one has*:

$$\| M_1 v \|_{B_1} \leqslant K \| M_2 v \|_{B_2}. \tag{2.2.2}$$

for any $v \in V$.

Denote by Ψ_0 the subspace of B_2^* formed by all functionals ψ orthogonal to $M_2(V)$. Consider the quotient space $\mathfrak{F} = B_2^*/\Psi_0$. As is known, this is a Banach space where the norm of an element, i.e. of an equivalence class $[\psi]$ represented by ψ, is defined in the following way:

$$\|[\psi]\|_{\mathfrak{F}} = \inf_{\psi_0 \in \Psi_0} \| \psi + \psi_0 \|_{B_2^*}.$$

As a by-product of the proof of Theorem 2.IV, one shows that, if (2.2.2) is satisfied and if ψ is a solution of (2.2.1), one has:

$$\|[\psi]\|_{\mathfrak{F}} \leqslant K \| \varphi \|_{B_1^*}. \tag{2.2.3}$$

(2.2.3) is called the *dual estimate formula* of (2.2.2).

It is self-evident that the existence principle 2.IV may be used only when the kernel V_2 of the transformation M_2 is contained in the kernel V_1 of M_1, in other words only if $M_2 v = 0$ implies $M_1 v = 0$.

When $V_2 \not\subset V_1$, a solution of (2.2.1) for any given φ cannot be obtained, since the following *necessary condition for the existence of* ψ must be satisfied:

$$\langle \varphi, M_1 v_2 \rangle = 0 \quad \text{for any } v_2 \in V_2. \tag{2.2.4}$$

However, the principle 2.IV may be reformulated in such a way as to include also the case $V_2 \not\subset V_1$. To this end, consider the quotient Banach space

$$Q = B_1/\overline{M_1(V_2)},$$

where, let us remember, the norm of an element (equivalence class) $[w]$ is given by

$$\|[w]\|_Q = \inf_{v_2 \in V_2} \| w + M_1 v_2 \|_{B_1}.$$

Let $\widetilde{M}_1$ be the linear transformation defined in V, with range in Q, which maps v_1 into $[M_1 v]$ of Q.

2.V. *A necessary and sufficient condition for the existence of a solution* ψ *of* (2.2.1), *for any given* φ *satisfying* (2.2.4), *is the existence of a constant* $K > 0$ *such that, for any* $v \in V$, *one has*:

$$\| \widetilde{M}_1 v \|_Q \leqslant K \| M_2 v \|_{B_2}. \tag{2.2.5}$$

As a matter of fact, this theorem, which appears at first glance as a generalization of Theorem 2.IV, may be easily inferred by this last one.[2]

[2] The first indication of Theorem 2.V has to be credited to Faedo [135]. However, Faedo does not make use of the concept of a quotient Banach space and gives a direct proof of 2.V which does not imply that the generalization of 2.V on 2.IV is only an apparent one.

For, if Φ is an element of the topological dual Q^* of Q, then it admits the following representation

$$\langle \Phi, [u] \rangle = \langle \varphi, u \rangle, \tag{2.2.6}$$

φ being an element of B_1^* orthogonal to $M_1(V_2)$. Conversely, if φ is a functional of such a type, then (2.2.6) defines an element of Q^*.

By (2.2.4) we can rewrite (2.2.1) as follows

$$\langle \Phi, \widetilde{M} v_1 \rangle = \langle \psi, M_2 v \rangle \qquad (v \in V),$$

Φ being defined by (2.2.6). Since $M_2 v = 0$ implies $\widetilde{M}_1 v = 0$, one may apply Theorem 2.IV and achieve the proof of 2.V.

Since $\| \Phi \|_{Q^*} = \| \varphi \|_{B_1^*}$, the dual inequality of (2.2.5) is always (2.2.3).

3. Examples of Dual Estimates

Let us consider, in the bounded and properly regular domain A of X^r, the second order linear differential operator with real coefficients, which (by the summation convention) may be written out in the following way:

$$Lv \equiv a^{ij}(x)\, v_{x_i x_j} + b^i(x) v_{x_i} + c(x) v \qquad (a^{ij} \equiv a^{ji});$$

Suppose $a^{ij}(x) \in C^2(\bar{A})$, $b^i(x) \in C^1(\bar{A})$ and $c(x)$ are Hölder continuous in $\bar{A}$ and never positive therein. Let $a^{ij}(x)\, \xi_i \xi_j > 0$ for $|\xi| > 0$. Set

$$L^* w = a^{ij} w_{x_i x_j} + b^{*i} w_{x_i} + c^* w$$

$$b^{*i} = 2\, a^{ij}_{x_j} - b^i, \qquad c^* = a^{ij}_{x_i x_j} + b^i_{x_i} + c.$$

Let $p \geqslant 1$ be a real number and δ an arbitrary positive number. For $v \in C^1(\bar{A}) \cap C^2(A)$ such that $Lv \in C^0(\bar{A})$ and for w enjoying analogous properties (replacing Lv by $L^* w$) one has:

$$\begin{aligned} &\int_A \{(v^2 + \delta)^{p/2} L^*(w) - w[p[(p-1)v^2 + \delta](v^2 + \delta)^{p/2-2} a^{hk} v_{x_h} v_{x_k} \\ &+ p(v^2 + \delta)^{p/2-1} v Lv + c(v^2 + \delta)^{p/2-1}[(1-p)v^2 + \delta]]\}\, dx \qquad (2.3.1) \\ &= \int_{\partial A} [p(v^2 + \delta)^{p/2-1} w v a^{hk} u_{x_h} \nu_k - (v^2 + \delta)^{p/2}(a^{hk} w_{x_h} \nu_k - b w)]\, d\sigma \end{aligned}$$

($\nu \equiv$ inward normal to ∂A; $b = b^i \nu_i - a^{ij}_{x_j} \nu_i$).

By (2.3.1), supposing $Lu = 0$ and taking as w a function such that $L^* w > 0$ in $\bar{A}$, $w = 0$ on ∂A, the following majorization is obtained, letting $\delta \to 0$

$$\left(\int_A |v|^p \,dx\right)^{1/p} \leqslant K_p \left(\int_{\partial A} |v|^p \,d\sigma\right)^{1/p}, \tag{2.3.2}$$

with

$$K_p = \left(\frac{\max_{\partial A} |a^{hk} w_h \nu_k|}{\min_{\bar{A}} |L^* w|}\right)^{1/p}. \tag{2.3.3}$$

Let V be the vector space formed by all functions v belonging to $C^1(\bar{A}) \cap C^2(A)$ which are solutions of the equation $Lv = 0$ in A. Let B_1 be the space $\mathfrak{L}^p(A)$ and B_2 the space $\mathfrak{L}^p(\partial A)$; let M_1 be the identity transformation and M_2 the transformation mapping any $v \in V$ into its trace on ∂A. For $p > 1$ (2.2.1) is written

$$\int_A v\varphi \,dx = \int_{\partial A} v\psi \,d\sigma \tag{2.3.4}$$

with $\varphi \in \mathfrak{L}^q(A)$ and $\psi \in \mathfrak{L}^q(\partial A)$ $\left(q = \frac{p}{p-1}\right)$, and, for $p = 1$,

$$\int_A v \,d\varphi = \int_{\partial A} v \,d\psi, \tag{2.3.5}$$

φ and ψ being two measures defined on the Borel sets of $\bar{A}$ and ∂A, respectively. On the other hand, by the known closure properties of the solutions of the equation $Lv = 0$, given φ, there exists a unique solution ψ of (2.3.4) [of (2.3.5)]. Assuming, however, $\varphi = L^*u$, with $u \in C^1(\bar{A}) \cap C^2(A)$, $L^*u \in C^0(\bar{A})$ and $u = 0$ on ∂A, one has:

$$\int_A vL^*u \,dx = -\int_A va^{hk} u_{x_h} \nu_k \,d\sigma$$

and, hence, by (2.3.4) $\psi = -a^{hk} u_{x_h} \nu_k$. The dual formula of (2.3.2) yields therefore:

$$\left(\int_{\partial A} |a^{hk} u_{x_h} \nu_k|^q \,d\sigma\right)^{1/q} \leqslant K_p \left(\int_A |L^*u|^q \,dx\right)^{1/q},$$

K_p being given by (2.3.3). If, on the contrary, (2.3.5) is used, one has to consider (2.3.2) for $p \to \infty$, which then yields the well known maximum principle: $\max_{\bar{A}} |v| = \max_{\partial A} |v|$. The dual formula is then the following:

$$\int_{\partial A} |a^{hk} u_{x_k} \nu_k| \,d\sigma \leqslant \int_A |L^*u| \,dx.$$

If the further condition $a^{hk} v_{x_h} \nu_k \equiv 0$ on $\partial_2 A$ is imposed on v, $\partial_2 A$ being a part of

∂A composed by regular hypersurfaces, putting $\partial_1 A = \partial A - \partial_2 A$, one has, as is known from second order elliptic equation theory,

$$\max_{\bar{A}} |v| \leqslant \max_{\partial_1 A} |v|. \tag{2.3.6}$$

The dual formula of (2.3.6) is the following:

$$\int_{\partial_1 A} |a^{hk} u_{x_h} \nu_k| \, d\sigma \leqslant \int_A |L^* u| \, dx,$$

which holds for any u satisfying the qualitative hypotheses specified above and, in addition, such that $u = 0$ on $\partial_1 A$, $a_{hk} u_{x_h} \nu_k = 0$ on $\partial_2 A$.

Formula (2.3.2) and, hence, its dual have been subsequently considered by Bramble and Payne [146], [147], who have determined the constant K_p in a more explicit way, without the introduction of functions, such as w satisfying particular conditions.

Let us consider other examples.

For a function v, enjoying the regularity properties specified above, the following estimate holds:

$$\left(\int_{\partial A} v^2 \, d\sigma \right)^{1/2} \leqslant K \left(\int_A [a^{hk} v_{x_h} v_{x_k} - cv^2] \, dx \right)^{1/2}, \tag{2.3.7}$$

where it has been assumed that

$$\frac{\partial}{\partial x_h} \left(a^{hk} \frac{\partial v}{\partial x_k} \right) + cv = 0 \tag{2.3.8}$$

in A and that $v(x^0) = 0$ in a fixed point x^0 of A, if $c \equiv 0$ in A.

For the proof of (2.3.7) and the explicit computation of K, the reader is referred to [148]. If now the role of (2.2.1) is played by the identity

$$\int_{\partial A} v a^{hk} u_{x_h} \nu_k \, d\sigma = - \int_A (a^{hk} u_{x_h} v_{x_k} - cuv) \, dx,$$

where u is a solution of (2.3.8), one ends up with the following dual formula of (2.3.7):

$$\left(\int_A (a^{hk} u_{x_h} u_{x_k} - cu^2) \, dx \right)^{1/2} \leqslant K \left(\int_{\partial A} |a^{hk} u_{x_h} \nu_k|^2 \, d\sigma \right)^{1/2}.$$

As a last example, let us infer the dual estimation formula concerning a fourth order equation. More precisely, let us consider the following estimate due to Miranda [149]:

$$\max_A |\operatorname{grad} v| \leqslant K \max_{\partial A} |\operatorname{grad} v| \tag{2.3.9}$$

relative to any solution of the biharmonic equation $\Delta_2\Delta_2 v = 0$, considered in a domain A of the plane x_1x_2, bounded by closed curves, having continuous tangents and curvatures.

Denoting by $\boldsymbol{u}$ a vector of components u_1, u_2, set:

$$L_i^*(u) = \Delta_2 u_i - \frac{\partial}{\partial x_i}\left(\frac{\partial u_1}{\partial x_1} + \frac{\partial u_2}{\partial x_2}\right)$$

$$M_i^*(u) = -\left(\frac{\partial u_1}{\partial x_1} + \frac{\partial u_2}{\partial x_2}\right)\frac{dx_i}{d\nu} + \frac{du_i}{d\nu} \qquad (i = 1, 2).$$

Within suitable assumptions on v and $\boldsymbol{u}$, one has, for any biharmonic function v in A and any vector $\boldsymbol{u}$ vanishing on ∂A,

$$\int_{\partial A}\left[\frac{\partial v}{\partial x_1} M_1^*(u) + \frac{\partial v}{\partial x_2} M_2^*(u)\right] d\sigma = -\int_A \left[\frac{\partial v}{\partial x_1} L_1^*(u) + \frac{\partial v}{\partial x_2} L_2^*(u)\right] dx_1\, dx_2.$$

In the present case such a relation plays the role of (2.2.1) and, hence, keeping in mind the fact that grad v describes a complete system in a suitable space of vectors continuous on ∂A, one obtains:

$$\int_{\partial A} (|M_1^*(u)|^2 + |M_2^*(u)|^2)^{1/2} d\sigma \leqslant K \int_A (|L_1^*(u)|^2 + |L_2^*(u)|^2)^{1/2} dx, \qquad (2.3.10)$$

valid for any u vanishing on ∂A. This is the dual formula of (2.3.9).

If (2.3.10) had been directly established, which it seems possible to do through procedures analogous to those to be described in the following Sections, one would have obtained (2.3.9). This suggests a way to prove (2.3.9) in X^r with $r > 2$.

An exposition of the results reviewed in this Section may be found in [150].

4. The Maximum Modulus Theorem for the Elasticity Equation

Among the applications of the procedures based upon the dual estimates, a particularly interesting one is represented by a *maximum modulus theorem* for the *three-dimensional* elasticity operator, which we shall now denote in the following way:

$$E(v) \equiv \Delta_2 v + k \text{ grad div } v \qquad (k > -1),$$

v being a 3-vector function.

The application consists in proving that, on suitable regularity assumptions on the domain A, one has for any solution of the equation $E(v) = 0$, continuous in $\bar{A}$,

$$\max_{\bar{A}} |v| \leqslant H \max_{\partial A} |v|, \qquad (2.4.1)$$

H being an *explicitly computable* constant depending only on A and k [151].

The achievement of this result requires the overcoming of considerable technical difficulties. We limit ourselves to the indication of the various steps of the proof and of the intermediate results obtained which are interesting by themselves.

Denote by Lw the following boundary operator (ν stands for the inward normal unit vector)

$$Lw = \left[\left(k - \frac{k}{k+2}\right) \operatorname{div} w\right] \nu + \left(1 + \frac{k}{2+k}\right) \frac{\partial w}{\partial \nu} + \frac{k}{2+k} (\nu \wedge \operatorname{rot} w).$$

In the following we shall denote by u a 3-vector function, belonging to $C^1(\bar{A}) \cap C^2(A)$, such that $Eu \in C^0(\bar{A})$ and satisfying the condition $u = 0$ on ∂A.

(2.4.1) is obtained by subsequently establishing the following results:

(1) Existence and computation of two positive constants H_1 and H_2, such that

$$\max_{\partial A} | Lu | \leqslant H_1 \max_{\bar{A}} | E(u) | + H_2 \left(\int_A | u |^2 dx \right)^{1/2}.$$

(2) Proof of the inequality

$$\left(\int_A | u |^2 dx \right)^{1/2} \leqslant H_3 \left(\int_A | E(u) |^2 dx \right)^{1/2}$$

with an explicit computation of H_3 [(3)] and, hence, existence and computation of H_4 such that:

$$\max_{\partial A} | Lu | \leqslant H_4 \max_{\bar{A}} | E(u) |.$$

(3) Computation of H_5 and H_6 such that

$$\max_{\bar{A}} | u(x) | \leqslant H_5 \max_{\partial A} | Lu | + H_6 \max_{\bar{A}} | E(u) |$$

and hence proof of the inequality

$$\max_{\bar{A}} | u(x) | \leqslant H_7 \max_{\bar{A}} | E(u) |. \tag{2.4.2}$$

(4) Deduction by duality from (2.4.2) of the inequality

$$\int_A | u | dx \leqslant H_7 \int_A | E(u) | dx.$$

(5) Existence and computation of two constants H_8 and H_9, such that

$$\int_{\partial A} | Lu | d\sigma \leqslant H_8 \int_A | E(u) | dx + H_9 \int_A | u | dx$$

---

(3) *See* the integral inequality on page 128.

and hence the deduction of the inequality

$$\int_A |Lu|\, d\sigma \leqslant H \int_A |E(u)|\, dx. \tag{2.4.3}$$

(6) Deduction of (2.4.1) by duality from (2.4.3).

To give an idea of the complication of this investigation, let us point out how the explicit computation of the constant H is achieved.

Suppose A to be such that its intersection with the spherical domain $|x - x^0| < \rho_A$ (x^0 being an arbitrary point of ∂A and ρ_A a number depending only on A) is a connected set Σ_{x^0}, which, assuming as the x_3 axis the normal direction to ∂A in x^0 and then taking x_1 and x_2 axes on the tangent plane through x^0 can be represented by the equation

$$x_3 = \varphi(x_1, x_2),$$

φ_{x_1} and φ_{x_2} being uniformly Hölder continuous (in the orthogonal projection of Σ_{x^0} onto the above-mentioned tangent plane), with coefficient c and exponent h.

Set:

$$M_1 = 4\pi c \left\{ \left[(1+\sqrt{2}c) \left(\frac{\sqrt{3}}{2} + \frac{|k|}{k+2} \right) + \frac{1}{\sqrt{2}} \frac{|k|}{k+2} \right] \frac{1}{2+h} + (1+\sqrt{2}c) \frac{2\sqrt{3}+3|k|}{k+2} \left(\frac{1}{h} + \frac{1}{h+1} \right) \right\},$$

$$N_1 = 4\pi c \left\{ \frac{1}{2} \frac{1}{2+h} \frac{\sqrt{3}(2+k)+(2+\sqrt{2})|k|}{2+k} + \frac{2\sqrt{3}+3|k|}{2+k} \left(\frac{1}{h} + \frac{1}{h+1} \right) \right\}$$

$$a_n = -4(2n+3)(2n+5)(2n+7)$$

$$\psi(t) = a_n (2t-1)^{2n+1}, \qquad F(t) = \int_1^t \frac{(t-\tau)^2}{2} \tau (1-\tau)^3 \psi(\tau)\, d\tau$$

$$b_n = \int_0^1 t F'''(t)\, dt = \frac{(2n+3)(2n+5)(2n+7)}{8(n+1)(n+2)(n+3)}$$

$$F_i = \max_{[0,1]} |F^{(i)}(t)|, \qquad \Phi_h = \max\{F_1, \ldots, F_h\}.$$

Let the positive integer n be chosen in such a way that

$$1 - \frac{|k|}{2+k} b_n > \frac{1}{2} \left(1 - \frac{|k|}{2+k} \right), \qquad M_1 |a_n| > 2\pi \left(1 - \frac{|k|}{2+k} b_n \right)$$

$$N_1 |a_n| > 2\pi \left(1 - \frac{|k|}{2+k} b_n \right).$$

Let $\epsilon > 0$ be such that

$$\epsilon \leqslant \left[\frac{2\pi\left(1-\frac{|k|}{2+k}b_n\right)}{M_1|a_n|}\right]^{1/h}\rho_A, \qquad \epsilon \leqslant \left[\frac{2\pi\left(1-\frac{|k|}{2+k}b_n\right)}{N_1|a_n|}\right]^{1/h}\rho_A.$$

Set:

$$M_2 = \frac{2\pi}{3}\,\frac{\sqrt{3}(k+2)+31\,|k|+18\sqrt{3}}{k+2}\,\Phi_3\,\epsilon,$$

$$M_3 = \sqrt{3\pi}\left(\sqrt{3}+|k|+\frac{\sqrt{2}\,|k|}{k+2}\right)F_5\,\epsilon^{-5/2}.$$

One has:

$$H_1 = \frac{M_2}{2\pi\left(1-\frac{|k|}{2+k}b_n\right)-M_1|a_n|\,\epsilon^h}, \qquad H_2 = \frac{M_3}{2\pi\left(1-\frac{|k|}{2+k}b_n\right)-M_1|a_n|\,\epsilon^h}$$

Moreover (*see* page 128), it turns out that:

$$H_3 = \frac{3\,d^2}{3+k},$$

d being the radius of a spherical domain containing A.

One has:

$$H_4 = H_1 + H_2 H_3\,(\mathrm{vol}\ A)^{1/2}.$$

Setting:

$$s_j^i(x,y) = \frac{1}{2}\,\delta_{ij}\Delta_2\,|x-y| - \frac{k}{2(1+k)}\,\frac{\partial^2|x-y|}{\partial x_i \partial x_j}, \qquad (i, j = 1, 2, 3)$$

one has:

$$H_5 = \frac{1}{4\pi}\max_{\bar{A}}\int_{\partial A}\left(\sum_{ij}|s_j^i(x,y)|^2\right)^{1/2} d\sigma_y,$$

$$H_6 = \frac{1}{4\pi}\max_{\bar{A}}\int_{A}\left(\sum_{ij}|s_j^i(x,y)|^2\right)^{1/2} dy,$$

$$H_7 = H_4 H_5 + H_6.$$

Let

$$N_2 = \left\{ \frac{10\sqrt{3}+(\sqrt{3}+14)\,|\,k\,|}{2\,(2+k)} \sup_{y\in A} \int_{\partial A} \left| \frac{\partial}{\partial \nu_x} \frac{1}{|\,x-y\,|} \right| d\sigma_x + \frac{|\,k\,|}{2+k}\,\epsilon^{-2}\,(\text{area }\partial A) \right\} |\,a_n\,|$$

$$N_3 = 3\left(\sqrt{3}+|\,k\,|+\frac{\sqrt{2}\,|\,k\,|}{2+k}\right) F_5\,\epsilon^{-3} \max_{y\in \bar{A}} \int_{\partial A} \frac{d\sigma_x}{|\,x-y\,|}\,.$$

One has:

$$H_8 = \frac{N_2}{2\,\pi\left(1-\dfrac{|\,k\,|}{2+k}\,b_n\right)-N_1\,|\,a_n|\,\epsilon^h}, \qquad H_9 = \frac{N_3}{2\,\pi\left(1-\dfrac{|\,k\,|}{2+k}\,b_n\right)-N_1\,|\,a_n|\,\epsilon^h},$$

and finally

$$H = H_8 + H_7 H_9,$$

i.e.

$$H = H_8 + H_1 H_5 H_9 + H_2 H_3 H_5 H_9 (\text{vol } A)^{1/2} + H_6 H_9.$$

5. Proof of a Conjecture of Mikhlin and Extension of the Marcel Riesz Inequalities to Three-Dimensional Space

If u is a two-variables harmonic function in a domain A, bounded by a closed curve Σ of class 2, the well known inequalities of Marcel Riesz state that, if $\partial u/\partial s$ and $\partial u/\partial \nu$ are the derivatives of u on Σ along the tangent direction and along the normal one, respectively, then assuming $u \in C^1(\bar{A})$, for $1 < p < +\infty$ one has:

$$\left(\int_\Sigma \left|\frac{\partial u}{\partial s}\right|^p ds\right)^{1/p} \leqslant K \left(\int_\Sigma \left|\frac{\partial u}{\partial \nu}\right|^p ds\right)^{1/p}, \tag{2.5.1}$$

$$\left(\int_\Sigma \left|\frac{\partial u}{\partial \nu}\right|^p ds\right)^{1/p} \leqslant K \left(\int_\Sigma \left|\frac{\partial u}{\partial s}\right|^p ds\right)^{1/p}. \tag{2.5.2}$$

Their extension to three-dimensional space would read:

$$\left(\int_\Sigma |\,\text{grad}_\Sigma\, u\,|^p\, d\sigma\right)^{1/p} \leqslant K \left(\int_\Sigma \left|\frac{\partial v}{\partial \nu}\right|^p d\sigma\right)^{1/p}, \tag{2.5.3}$$

$$\left(\int_\Sigma \left|\frac{\partial u}{\partial \nu}\right|^p d\sigma\right)^{1/p} \leqslant K \left(\int_\Sigma |\,\text{grad}_\Sigma\, u\,|^p\, d\sigma\right)^{1/p}. \tag{2.5.4}$$

Here $\mathrm{grad}_\Sigma u$ stands for the 'tangent gradient' of u on Σ.

While (2.5.1) and (2.5.2) are the same inequality (it is enough to replace u by its conjugate harmonic in (2.5.2)), this is not true for (2.5.3) and (2.5.4). Mikhlin in [152] proves (2.5.3), but he limits himself to a conjecture about the possibility of proving (2.5.4). This has been achieved by De Vito [153], using the procedures and techniques described in the former Sections of this Chapter. He assumes $\Sigma \in C^{2,h}$.

Let $H_1^p(\Sigma)$ $(1 < p < +\infty)$ be the space of the functions v defined on Σ, such that $v \in \mathfrak{L}^p(\Sigma)$, $\mathrm{grad}_\Sigma v \in \mathfrak{L}^p(\Sigma)$ with the norm

$$\| v \|_p = \left(\int_\Sigma | v |^p d\sigma \right)^{1/p} + \left(\int_\Sigma | \mathrm{grad}_\Sigma v |^p d\sigma \right)^{1/p}.$$

The treatment of [153] rests on the theorem stating that the transformation

$$F\varphi = \int_\Sigma \varphi(x) \frac{\partial}{\partial \nu_x} \frac{1}{| x - y |} d\sigma_x + 2\pi\varphi(y),$$

maps $H_1^p(\Sigma)$ onto itself one-to-one. Such a result is based upon a preliminary lemma, which states that the harmonic function u, represented by the double layer

$$u(y) = \int_\Sigma \varphi(x) \frac{\partial}{\partial \nu_x} \frac{1}{| x - y |} d\sigma_x, \tag{2.5.5}$$

is such that

$$\| \mathrm{grad}_\Sigma u \|_{\mathfrak{L}^p(\Sigma)} = \mathcal{O}(\| \mathrm{grad}_\Sigma \varphi \|_{\mathfrak{L}^p(\Sigma)}). \tag{2.5.6}$$

It follows, through the above-mentioned theorem, that if $u \in H_1^p(\Sigma)$ and $F\varphi = u$, then it turns out that

$$\| \varphi \|_p = \mathcal{O}(\| u \|_p). \tag{2.5.7}$$

On the other hand, for the function u given by (2.5.5), one may prove that

$$\left\| \frac{\partial u}{\partial \nu} \right\|_{\mathfrak{L}^p(\Sigma)} = \mathcal{O}(\| \varphi \|_p). \tag{2.5.8}$$

Therefore (2.5.4) is easily obtained by combining (2.5.8) and (2.5.7).

De Vitó also constructs some examples showing that (2.5.4) is false for $p = 1$ and for $p = \infty$.

6. The Estimates for the Exterior Differential Forms Equivalent to the Existence Theorems of De Rham

The general existence principle considered in Section 2 reduces an existence theorem of linear analysis to the proof of an 'a priori' estimate. From this point of view the writer looked in [125] for the estimates related to the well known existence theorems of De Rham, for the exterior differential forms on a compact and orientable differential manifold V^r. This allowed a proof of those theorems of purely analytical nature.

Let $B_1, \ldots, B_m$ be Borel sets covering V^r. Let B_s $(s = 1, \ldots, m)$ be contained, together with its closure $\bar{B}_s$, in an admissible map of V^r, where there exists, therefore, a local coordinate system $x^1, \ldots, x^r$. Let u be a continuous k-form on V^r and let

$$u = \frac{1}{k!} \sum_{s_1 \ldots s_k} u_{s_1 \ldots s_k} \, dx^{s_1} \cdots dx^{s_k}$$

be the local representation of u in the considered map.

For $p \geqslant 1$ assume:

$$\| u \|_p = \sum_{s=1}^{m} \sum_{i_1 < \ldots < i_k} \left(\int_{B_s} | u_{i_1 \ldots i_k} |^p \, dx^1 \cdots dx^r \right)^{1/p}, \tag{2.6.1}$$

where for $p = \infty$ the right-hand side becomes

$$\sum_{s=1}^{m} \sum_{i_1 < \ldots < i_k} \underset{B_s}{\text{pseudo-sup}} \, | u_{i_1 \ldots i_k} |.$$

The space, obtained by means of functional completion with respect to the norm (2.6.1), is the $\mathfrak{L}^p$ space of the k-forms.

If a different covering $B'_1, \ldots, B'_n$ of V^r is considered, a norm $\| u \|'_p$ is obtained, which is isomorphic to $\| u \|_p$ and hence defines, up to a Banach isomorphism, the same $\mathfrak{L}^p$ space of the k-forms.

Now the estimate allowing us, through the general existence principle of Section 2, to achieve the proof of the De Rham theorems is expressed by the following theorem:

2.VI. *Given the k-form v of class* C^1 *on* V^r, *let* $\mathfrak{V}_0$ *be the manifold of the closed (i.e. such that* $dv_0 = 0$*) k-forms* v_0 *of class* C^1 *on* V^r. *For any* $p > r$ *there exists a constant* $K_p > 0$, *depending only on* V^r *and p, such that*

$$\inf_{\mathfrak{V}_0} \| v + v_0 \|_\infty \leqslant K_p \| dv \|_p . \tag{2.6.2}$$

An analogous estimate of (2.6.2) which also allows us to develop in a very wide contex the theory of the exterior differential forms on V^r, is the following one:

$$\inf_{\mathfrak{V}_0} \| v + v_0 \|_2 \leqslant K \| dv \|_2 ,$$

which represents an extension to the k-forms of the well known Poincaré inequality for the scalar functions.

7. Boundary Value Problems for Second Order Elliptic-Parabolic Equations and Related Estimates

In the framework of the investigations originated by the applications of the general existence principle of Section 2, are included those, begun by the writer in 1956, on the 'well posed' boundary problems for a general second order elliptic-parabolic partial differential equation. Let A be a domain (open set) of the cartesian space X^r and consider therein the second order linear differential equation in its more general form:

$$\sum_{h,k}^{1,r} a_{hk} \frac{\partial^2 u}{\partial x_h \partial x_k} + \sum_{h=1}^{r} b_h \frac{\partial u}{\partial x_h} + cu = f \qquad (a_{hk} = a_{kh}). \tag{2.7.1}$$

a_{hk}, b_h, c, f are given real functions, defined in A.

The differential operator on the left-hand side of (2.7.1) will be denoted concisely by $\mathcal{E}(u)$. We shall consider the case of $\mathcal{E}(u)$ *positive* (or *non-negative*) *elliptic-parabolic* at any point of A, i.e. we shall suppose that, for any fixed point $x \equiv (x_1, x_2, \ldots, x_r)$, the quadratic form in the real variables $\lambda_1, \lambda_2, \ldots, \lambda_r$:

$$\Phi \equiv \sum_{h,k}^{1,r} a_{hk}(x) \lambda_h \lambda_k$$

is positive definite or semidefinite, the identically zero case being not excluded.

As is known, (2.7.1) is said to be *totally elliptic* if the quadratic form Φ is positive definite for any $x \in A$, *totally parabolic* in A if Φ is positive semidefinite for any such x and of *mixed type* (*elliptic-parabolic*) in other cases. We comment that first order linear equations

$$\sum_{h=1}^{r} b_h \frac{\partial u}{\partial x_h} + cu = f$$

are included as particular cases into the second order totally parabolic ones.

Usually it is understood that an *integration problem in the large* relative to (2.7.1) is a problem consisting of the determination of a solution of (2.7.1) in the domain A, which satisfies certain possible further side conditions.

If the domain A is bounded, as will be assumed from now on, these further conditions are, usually, *boundary conditions*, i.e. denoting by Σ the boundary of A, they amount to prescribing that a well determined combination of u and its first partial derivatives, should coincide with a known function on a given set, contained in Σ.

In this Section we will only consider the case in which the boundary condition consists of prescribing the value of u on a part of Σ or on the whole of Σ, according to the different cases.[4]

It is known that, if (2.7.1) is totally elliptic, the boundary condition amounts to giving u on the whole of Σ, thus obtaining the classical Dirichlet problem for (2.7.1), which,

---

(4) For an investigation relative to more general boundary conditions, *see* [155], [156].

under suitable smoothness assumptions on the coefficients, on the known function f in (2.7.1) and on the domain A, is a *well posed boundary problem.*

For the totally elliptic case, therefore, a complete generality is allowed for the equation in formulating the boundary problem for (2.7.1), except for the above-mentioned smoothness assumptions. This is not the case for the remaining situations as, for instance, the totally parabolic one for which boundary problems relative to very particular types of equations are considered. More precisely the parabolic case usually amounts to considering the particular equation

$$\sum_{h,k}^{1,r-1} a_{hk} \frac{\partial^2 u}{\partial x_h \partial x_k} + \sum_{h=1}^{r-1} b_h \frac{\partial u}{\partial x_h} + cu - \frac{\partial u}{\partial x_r} = f \tag{2.7.2}$$

and, supposing Σ to be smooth enough, in assigning u on that part Σ^* of Σ, where the tangent hyperplane is not orthogonal to the x_r axis or, in the opposite case, where the inward normal to Σ^* at the points of A has the same direction as x_r. The set of such points of Σ^* is usually assumed to be composed of parts of a hyperplane. If one admits that the operator in the $(r-1)$ variables $x_1, x_2, \ldots, x_{r-1}$

$$\mathcal{E}_0(u) \equiv \sum_{h,k}^{1,r-1} a_{hk} \frac{\partial^2 u}{\partial x_h \partial x_k} + \sum_{h=1}^{r-1} b_h \frac{\partial u}{\partial x_h} + cu$$

is totally elliptic for any fixed x_r, the boundary problem is well posed, in a sense to be specified below.

For $r = 2$ we have the classical heat equation case:

$$\frac{\partial^2 u}{\partial x_1^2} - \frac{\partial u}{\partial x_2} = f(x_1, x_2), \tag{2.7.3}$$

whose typical boundary problem consists in assuming as A a domain defined by the conditions

$$f_1(x_2) < x_1 < f_2(x_2), \qquad 0 < x_2 < a,$$

$f_1(x_2)$ and $f_2(x_2)$ being two functions continuous together with their first derivatives in the closed interval $(0, a)$, and such that $f_1(x_2) < f_2(x_2)$ there. Denoting by C_i the curve whose equation is $x_1 = f_i(x_2)$ $(i = 1, 2)$ and by T_1 the segment $x_2 = 0$, $f_1(0) < x_1 < f_2(0)$ one prescribes to the unknown u the values on $C_1 \cup T_1 \cup C_2$.

Such problems for equations (2.7.2) and, in particular, (2.7.3), exhibit an essential difference from the totally elliptic case, *since one cannot assign in an arbitrary way the values of u on the whole of Σ, but only on a part of it.*

Consider the very simple totally parabolic equation

$$\frac{\partial u}{\partial x_1} = f(x_1, x_2) \tag{2.7.4}$$

In this case, too, one cannot assign the values of u on the whole of Σ, but u is determined, as it is easily seen, as soon as it is known on that part of Σ, in any point of which the inward normal n to A forms an obtuse angle with the x_1 axis (continuous line in Fig. 2.1) or on that part of Σ where such an angle is acute.

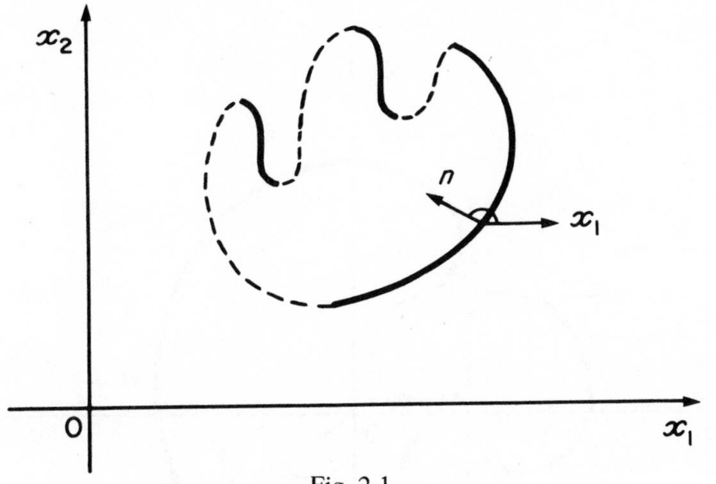

Fig. 2.1.

If, on the contrary, the following equation is considered:

$$x_1 \frac{\partial u}{\partial x_1} + x_2 \frac{\partial u}{\partial x_2} + cu = f(x_1, x_2) \tag{2.7.5}$$

c being a continuous negative function, one can prove that u, under suitable smoothness assumptions, turns out to be determined as soon as it is determined on that part of Σ, at any point of which the inward normal axis to A forms an obtuse angle with the outward drawn vector from the origin (continuous curve in Fig. 2.2).

Such a fact, when c is constant, is a simple consequence of the characteristic property of homogeneous functions.

There exist first order equations which, considered in particular domains, are such that *the unknown u is determined only by the knowledge of the known term*, i.e. the value of u can be arbitrarily prescribed in no part of Σ. Such a situation was explicitly pointed out as early as 1928 by Picone [157], who proved that, if one considers the equation

$$b_1(x_1, x_2) \frac{\partial u}{\partial x_1} + b_2(x_1, x_2) \frac{\partial u}{\partial x_2} + c(x_1, x_2)\, u = 0 \tag{2.7.6}$$

in the rectangle $R = A \cup \Sigma$:

$$x_1' \leqslant x_1 \leqslant x_1''; \qquad x_2' \leqslant x_2 \leqslant x_2'', \tag{2.7.7}$$

and if the continuous functions b_1, b_2, c satisfy the conditions

$$b_1(x_1', x_2) \geqslant 0, b_1(x_1'', x_2) \leqslant 0, b_2(x_1, x_2') \geqslant 0, b_2(x_1, x_2'') \leqslant 0; \tag{2.7.8}$$

$$c(x_1, x_2) < 0 \qquad \text{in } A \cup \Sigma, \tag{2.7.9}$$

then the unique solution of (2.7.7) in the whole of A is the identically zero function.

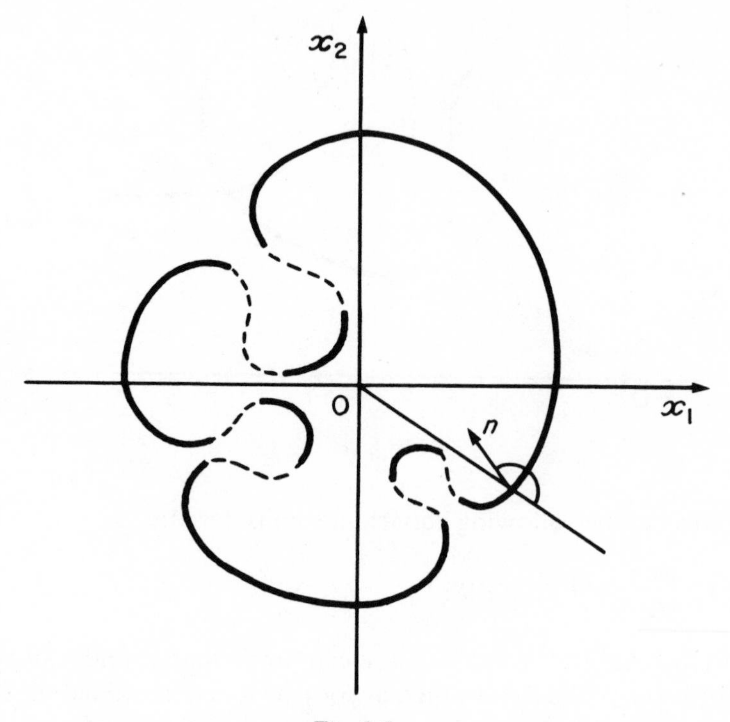

Fig. 2.2.

The examples we have considered show the essential difference between various particular cases of (2.7.1) so far examined, as far as the problem is concerned comprising the determination of that part of Σ, where the function u can be prescribed.

Therefore in [155] and [156] the following problem has been treated: *if the positive elliptic-parabolic equation* (2.7.1), *in its more general form, is considered, determine a part* Σ^* *of* Σ *such that the boundary problem*

$$\mathcal{E}(u) = f \qquad \text{in } A, \quad (2.7.10) \qquad u = \varphi \qquad \text{on } \Sigma^* \quad (2.7.11)$$

is well posed.

At this point it must be stated in a precise way what is meant by a 'well posed boundary problem'.

After Hadamard, we shall say that problem (2.7.10), (2.7.11) is well posed if it is possible to prescribe in advance a class $\mathfrak{A}$ of functions, such that the operator $\mathcal{E}(u)$, as well as the boundary condition (2.7.11), has a well defined meaning for any function u belonging to $\mathfrak{A}$, in such a way that

(a) there exists the solution of the problem in $\mathfrak{A}$, no matter how f and φ are assigned, satisfying at most some smoothness assumptions (continuity, Hölder continuity, integrability and so on);

(b) such a solution is unique;

(c) it depends 'continuously' on the data of the problem.

Point (c) requires a further comment about the meaning of the sentence '*the solution depends continuously on the data of the problem*'. This means that $\mathfrak{U}$ as well as the manifold $\mathfrak{F}$, spanned by the pair of functions f and φ, are to be considered as function spaces, where a topology or, in particular, a metric has been introduced and that, considering the transformation which maps an arbitrary element of $\mathfrak{F}$ into the solution of problem (2.7.10), (2.7.11), in the class $\mathfrak{U}$ this is continuous with respect to the topologies introduced in $\mathfrak{F}$ and in $\mathfrak{U}$.

The domain A of X^r will be assumed bounded, with boundary Σ formed by a finite number of regular hypersurface pieces of class 1, such that an arbitrary pair of them has common points at most on their boundaries. Furthermore let the *closed domain* $A \cup \Sigma$ have Σ as its boundary and let it be *regular*, in the sense that the classical Gauss–Green formulae, transforming hypersurface integrals extended to Σ into integrals extended to A, hold there.

Let the coefficients a_{hk} of (2.7.1) belong to the class 2 in $A \cup \Sigma$, the b_h be of class 1 and c be continuous there. As already mentioned, $\mathscr{E}(u)$ will be supposed positive elliptic-parabolic in A.

Let us denote by $\Sigma_1, \Sigma_2, \ldots, \Sigma_p$ the regular hypersurface portions composing Σ. Let Σ_i' be the set of the points of Σ_i in which the tangent hyperplane to Σ_i has characteristic slope with respect to the equation (2.7.1).[5]

In any point of Σ_i not belonging to the border of Σ_i the function

$$b(x) = \sum_{h=1}^{r} \left(b_h - \sum_{k=1}^{r} \frac{\partial a_{hk}}{\partial x_k} \right) \cos (x_h, n),$$

is well defined. Here $\cos (x_h, n)$ stands for the direction cosine, with respect to the coordinate axis x_h, of the axis n normal to Σ_i, pointing towards the interior of A. $b(x)$ may be considered, by continuity, to be defined on the whole of Σ_i.

Let $\Sigma_i^{(1)}$ be the set of the points of Σ_i', for which $b \geqslant 0$ and let us set

$$\Sigma^{(1)} = \Sigma_1^{(1)} \cup \Sigma_2^{(1)} \cup \ldots \cup \Sigma_p^{(1)},$$

$$\Sigma^{(2)} = (\Sigma_1' \cup \Sigma_2' \cup \ldots \cup \Sigma_p') - \Sigma^{(1)},$$

$$\Sigma^{(3)} = \Sigma - (\Sigma^{(1)} \cup \Sigma^{(2)}).$$

The set Σ is, thus, decomposed as the sum of the pairwise disjoint sets $\Sigma^{(1)}, \Sigma^{(2)}, \Sigma^{(3)}$. Let us make the explicit remark that some of these sets may also be empty. Such sets are invariant with respect to regular coordinate transformations in X^r.

---

[5] As is known, an hypersurface of S_r is said to have a characteristic slope at the point x, with respect to (2.7.1), if the direction cosines of its normal n satisfy the equation

$$\sum_{h,k}^{1,r} a_{hk}(x) \cos (x_h, n) \cos (x_k, n) = 0.$$

The general boundary value problem (2.7.10), (2.7.11), representing the aim of the present Section, is

(G) *Let the function* $\varphi(x)$ *be given on that part* Σ^* *of* Σ *defined in the following way*: $\Sigma^* = \Sigma^{(2)} \cup \Sigma^{(3)}$. *Determine a solution* u *of* (2.7.1) *satisfying the boundary condition* (2.7.11).

Before examining the existence and uniqueness questions, which will lead to the consideration of the functional class $\mathfrak{A}$, to which the unknown u is required to belong, let us remark that all the above-mentioned particular cases, concerning elliptic or parabolic equations, are included in problem (G).

If (2.7.1) is totally elliptic, it turns out that $\Sigma^{(3)} \equiv \Sigma \equiv \Sigma^*$, so that problem (G) reduces to the classical Dirichlet one for such equations. When (2.7.1) reduces to the totally parabolic equation (2.7.2) and $\mathcal{E}_0(u)$ is, for any fixed x_r, totally elliptic in the remaining variables, it is easily seen that $\Sigma^{(1)}$ is formed by those possible parts of Σ at any point of which the normal axis n is parallel and opposite to x_r, while $\Sigma^{(2)}$ is formed by the points of Σ, on which the normal axis is parallel to x_r, with the same direction. At any point of $\Sigma^{(3)}$, n forms with the x_r axis an angle different from 0 or π. Problem (G) is thus recognized to be the one already discussed for (2.7.2). If, however, the assumption made on $\mathcal{E}_0(u)$ does not hold and this operator is, for instance, positive parabolic in $x_1, \ldots, x_{r-1}$, the set Σ^* is in this case a proper subset of the set Σ^* relative to the case just considered.

Let us consider as an example the equation:

$$\frac{\partial^2 u}{\partial x_1^2} - 2\lambda \frac{\partial^2 u}{\partial x_1 \partial x_2} + \frac{\partial^2 u}{\partial x_2^2} - \frac{\partial u}{\partial x_3} - u = f(x_1, x_2, x_3),$$

λ being a constant such that $0 \leqslant \lambda \leqslant 1$.

Let A be the domain of X_3 having the form of an orthogonal triangular prism, bounded by the planes $x_3 = 0, x_3 = 1, x_1 = 0, x_2 = 0, x_1 + x_2 - 1 = 0$.

It turns out that if $0 \leqslant \lambda < 1$, then $\Sigma^{(1)}$ is formed by the face DCE (*see* Fig. 2.3), $\Sigma^{(2)}$ by ABO and $\Sigma^{(3)}$ by ABCD $\cup$ BCEO $\cup$ OEDA.

If, on the contrary, $\lambda = 1$, one has: $\Sigma^{(1)} =$ DCE $\cup$ ABCD, $\Sigma^{(2)} =$ ABO, $\Sigma^{(3)} =$ BCEO $\cup$ OEDA. So that, while for $0 \leqslant \lambda < 1$ u is assigned on ABO $\cup$ ABCD $\cup$ BCEO $\cup$ OEDA, for $\lambda = 1$, u is assigned only on ABO $\cup$ BCEO $\cup$ OEDA.

For the case of equation (2.7.4), $\Sigma^{(3)}$ being empty, *which is true for any first order equation*, it turns out that $\Sigma^* = \Sigma^{(2)}$, and in this way all points of Σ where the inward normal to A forms an obtuse angle with the x_1 axis, are obtained.

In an analogous way it is seen that, for equation (2.7.5), $\Sigma^* \equiv \Sigma^{(2)}$ is formed by all points of Σ, where the inward normal to A forms an obtuse angle with the outward drawn vector from the origin. For equation (2.7.6) the set $\Sigma^* \equiv \Sigma^{(2)}$ is empty, if conditions (2.7.8) are satisfied in the rectangle defined by (2.7.7).

Since the uniqueness of the solution of problem (G) in a suitable class is ensured, as we shall see in a moment, if a condition such as (2.7.9) holds, an explanation is achieved for the phenomenon discovered by Picone for (2.7.6).

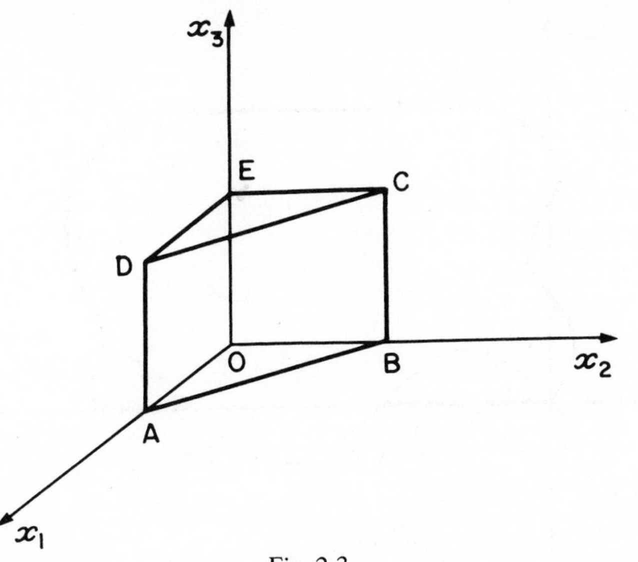

Fig. 2.3.

Let us consider some additional particular cases of problem (G), concerning totally parabolic equations.

Let $\mathcal{E}_0(u)$ be the operator introduced above, supposed to be positive elliptic, for any fixed x_r, in the variables $x_1, \ldots, x_{r-1}$. Assume that (2.7.1) reduces to the following equation:

$$\mathcal{E}(u) \equiv \mathcal{E}_0(u) + b_r(x)\ \frac{\partial u}{\partial x_r} = f, \tag{2.7.12}$$

$b_r(x)$ being a function of variable sign in $A \cup \Sigma$.

Denoting by $W^{(+)}\,[W^{(-)}]$ the possible part of Σ, in any point of which the normal axis n is parallel to the axis x_r with the same (opposite) direction and denoting by $W_1^{(+)}\,[W_1^{(-)}]$ the set of the points of $W^{(+)}\,[W^{(-)}]$ where $b_r(x) \geqslant 0$ ($b_r(x) \leqslant 0$) and by $W_2^{(+)}\,[W_2^{(-)}]$ the remaining part, one has:

$$\Sigma^{(1)} = W_1^{(+)} \cup W_1^{(-)},$$

$$\Sigma^{(2)} = W_2^{(+)} \cup W_2^{(-)}.$$

For instance the two-variables equation

$$\frac{\partial^2 u}{\partial x_1^2} - \sin x_1\ \frac{\partial u}{\partial x_2} = f(x_1, x_2).$$

is included in (2.7.12).

If we consider the domain A (normal with respect to the x_2 axis) of Fig.2.4, the part $\Sigma^{(1)}$ of Σ is the light-lined one. The $\Sigma^* = \Sigma^{(2)} \cup \Sigma^{(3)}$ is the heavy-lined one, $\Sigma^{(2)}$ being in particular formed by all segments of Σ^* parallel to the x_1 axis.

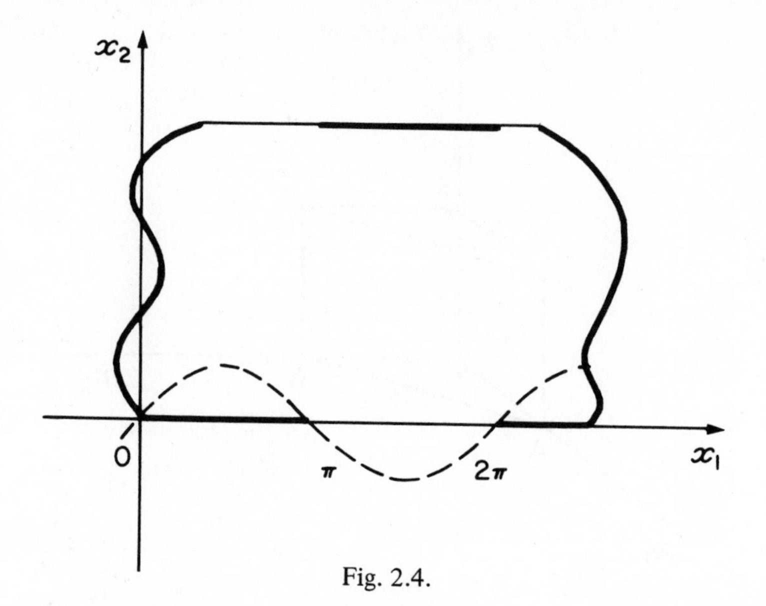

Fig. 2.4.

The totally parabolic equation in two variables

$$(x_1^2 + x_2^2 - 1)\frac{\partial^2 u}{\partial x_1^2} + x_2(x_1^2 + 2x_2^2 - 2)\frac{\partial u}{\partial x_2} + c(x_1, x_2)\,u = f(x_1, x_2) \qquad (2.7.13)$$

is not included in (2.7.12). Consider (2.7.13) in the domain A of the x_1x_2 plane determined by the conditions

$$x_1^2 + x_2^2 > 1, \qquad 0 < x_1 < 2, \qquad -1 < x_2 < 1.$$

The parts $\Sigma^{(2)}$ and $\Sigma^{(3)}$ composing the boundary are shown in Fig. 2.5. In this case $\Sigma^{(1)}$ consists only of the two points (0, 1) and (0, −1). If (2.7.13) is considered in the domain A′, formed only by the points of A having positive ordinate, the part $\Sigma^{(1)}$ of the boundary of A′ is formed by the segment (1.2) of the x_1 axis and by the point (0.1).

Let us consider, as a last example, a *totally parabolic equation, which does not reduce to a first order equation and which exhibits a peculiarity analogous to that remarked by Picone for* (2.7.6).

The equation is

$$x_2^2\frac{\partial^2 u}{\partial x_1^2} - 2x_1x_2\frac{\partial^2 u}{\partial x_1 \partial x_2} + x_1^2\frac{\partial^2 u}{\partial x_2^2} - 2x_1\frac{\partial u}{\partial x_1} - 2x_2\frac{\partial u}{\partial x_2} + cu = f(x_1, x_2). \qquad (2.7.14)$$

If no matter what circle in the x_1x_2 plane, centred on the origin, is taken as the domain A, since the boundary of A is a characteristic curve for (2.7.14) and since it turns out that

$$b(x) = -x_1\cos(x_1, n) - x_2\cos(x_2, n) = |\sqrt{x_1^2 + x_2^2}|,$$

it is seen that Σ^* is empty. In the starlike domain shown in Fig. 2.6, on the contrary, Σ^* coincides with the whole boundary (dashed line in Fig. 2.6).

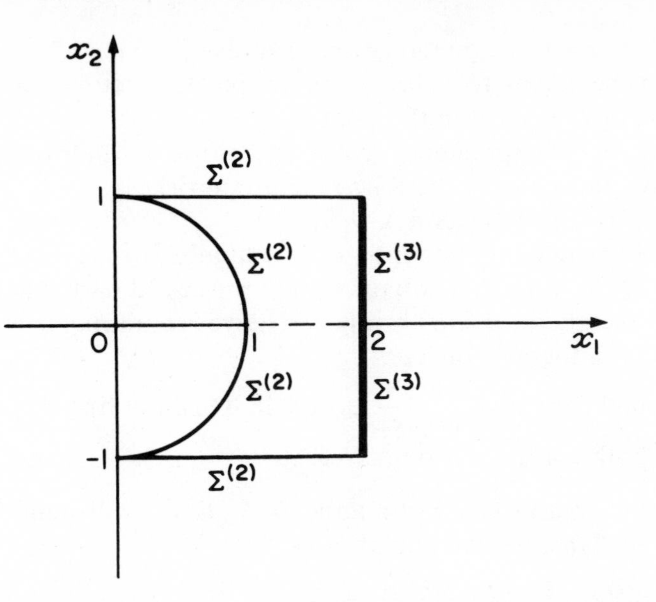

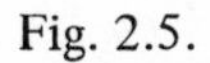
Fig. 2.5.

Since the condition $c < 0$, as we shall see later on, implies the uniqueness of the solution of problem (G) within a suitable class, any suitably smooth solution of (2.7.14) for $f \equiv 0$ in a domain A, containing the origin (for instance the starlike one just indicated), will necessarily vanish on the largest circular domain centred on the origin and contained in A.

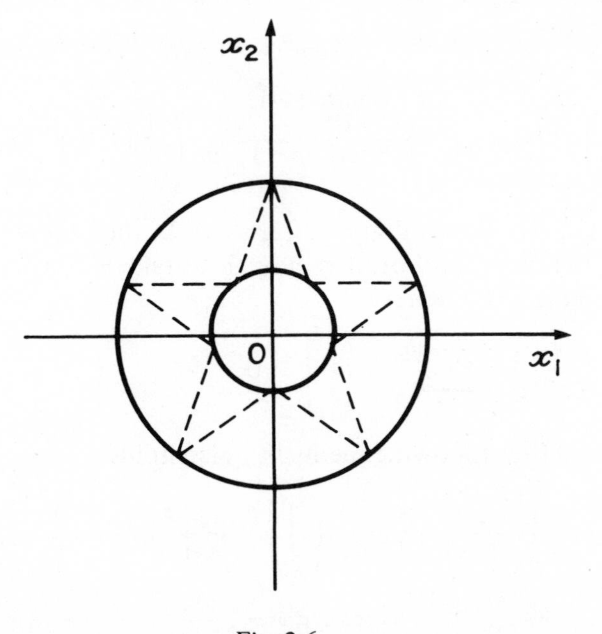

Fig. 2.6.

Let us point out now some majorization formulae [155], [156]. Such estimates hold for functions belonging to a class $\mathcal{C}$ to be specified below. These estimates lead to the uniqueness theorem within the class $\mathcal{C}$.

Let us consider, in correspondence to the operator $\mathcal{E}(u)$ and to the domain A, the class $\mathcal{C}$ of the functions having the following properties:

(a) any $u \in \mathcal{C}$ is continuous on $A \cup \Sigma$;

(b) $A \cup \Sigma$ may be decomposed into a finite number of regular domains $T_1, T_2, \ldots, T_s$ such that, whatever first or second partial derivative of u occurs in the operator $\mathcal{E}(u)$(6) this derivative is continuous in any internal point of T_k and is integrable on T_k;

(c) if the second derivative $\dfrac{\partial^2 u}{\partial x_h \partial x_k}$ occurs in $\mathcal{E}(u)$, the first derivatives $\dfrac{\partial u}{\partial x_h}$ and $\dfrac{\partial u}{\partial x_k}$ are both continuous on $A \cup \Sigma$.

Denote by p an arbitrary real number $\geqslant 1$. $\mathcal{E}^*(u)$ will stand for the *adjoint operator* of $\mathcal{E}(u)$, i.e. the operator

$$\mathcal{E}^*(u) = \sum_{h,k}^{1,r} \frac{\partial^2 (a_{hk} u)}{\partial x_h \partial x_k} - \sum_{h=1}^{r} \frac{\partial (b_h u)}{\partial x_h} + cu.$$

By referring (a), (b), (c) to the operator $\mathcal{E}^*(u)$ we define the class $\mathcal{C}^*$.

The following theorem holds:

2.VII. *If there exists, for a fixed p, a function w, of class two in* $A \cup \Sigma$, *satisfying, for any u belonging to* $\mathcal{C}$, *the conditions*

$$w \leqslant 0 \qquad \text{in } A \cup \Sigma, \qquad \mathcal{E}^*(w) + (p-1)\,cw > 0 \qquad \text{in } A \cup \Sigma$$

and vanishing on Σ^* *and such that* $|\mathcal{E}(u)|^p$ *is integrable on* A, *one has*:

$$\left(\int_A |u|^p \, dx \right)^{1/p} \leqslant p \, \frac{\max\limits_{A \cup \Sigma} |w|}{\min\limits_{A \cup \Sigma} (\mathcal{E}^*(w) + (p-1) cw)} \left(\int_A |\mathcal{E}(u)|^p \, dx \right)^{1/p}.$$

We remark that, if it happens that $c < 0$ in $A \cup \Sigma$, then there exist p and w, satisfying the conditions of the theorem. For, it is enough to take $w = -1$ and as p whatever real number such that in $A \cup \Sigma$ we have

$$\min_{A \cup \Sigma} \left[-\sum_{h,k}^{1,r} \frac{\partial^2 a_{hk}}{\partial x_h \partial x_k} + \sum_{h=1}^{r} \frac{\partial b_h}{\partial x_h} - pc \right] > 0.$$

For the case $c < 0$, the following inequality also holds:

$$\left(\int_A |u|^p |c| \, dx \right)^{1/p} \leqslant \frac{p}{p+1-p_0} \left(\int_A \left| \frac{\mathcal{E}(u)}{c} \right|^p |c| \, dx \right)^{1/p} \tag{2.7.15}$$

(6) We say that a first derivative $\dfrac{\partial u}{\partial x_h}$ $\left(\text{or a second derivative } \dfrac{\partial^2 u}{\partial x_h \partial x_k}\right)$ of u occurs in the operator (u), if the function b_h (or the function a_{hk}) does not vanish identically.

where p_0 is the smallest positive integer such that

$$p_0 |c| > |c| + \sum_{h,k}^{1,r} \frac{\partial^2 a_{hk}}{\partial x_h \partial x_k} - \sum_{h=1}^{r} \frac{\partial b_h}{\partial x_h}.$$

(2.7.15) holds for $p \geqslant p_0$. For $p \to \infty$ it yields, through a known lemma of F. Riesz, when $\mathcal{E}(u)$ is bounded on $A \cup \Sigma$,

$$\max_{A \cup \Sigma} |u| \leqslant \sup_{A} \left|\frac{\mathcal{E}(u)}{c}\right|. \text{[7]} \qquad (2.7.16)$$

The uniqueness theorems for problem (G), within the class $\mathcal{C}$ inferred by the above estimates, may be formulated immediately.

Let us finally notice the following theorem, which represents the maximum principle for the general equation (2.7.1), when $f \equiv 0$.

2. VIII. *If* $c < 0$ *in* $\bar{A}$, *then for any solution* $u \in \mathcal{C}$ *in* A *of the equation* $\mathcal{E}(u) = 0$, *one has*:

$$|u(x)| < \max_{\bar{\Sigma}^*} |u(\xi)| \qquad \text{for } x \in A.$$

Through this inequality and (2.7.16) it is possible to infer, if $c < 0$ in $\bar{A}$, the following inequality for any solution of problem (G) belonging to $\mathcal{C}$:

$$\max_{\bar{A}} |u| \leqslant \max \left(\sup_{\Sigma^*} |\varphi|, \sup_{A} \left|\frac{f}{c}\right| \right).$$

It may be verified, by means of particularly simple examples, that, in general, there is no existence theorem for problem (G), considered in the class $\mathcal{C}$, in connection with the uniqueness theorem within the class $\mathcal{C}$. To this end, it is enough to consider (2.7.4) or (2.7.5), c being a negative constant, in the domains shown in Figs. 2.1 and 2.2, respectively. It is immediately seen that problem (G), for the above-mentioned equations in these domains, does not admit, in general, a solution belonging to $\mathcal{C}$ and, hence, in particular, continuous.

Analogous examples may be produced for second order equations. In this case a wider functional class $\mathfrak{A}$ has to be considered, where the operator $\mathcal{E}(u)$ has to be *extended* and, once a meaning is assigned to the boundary condition (2.7.11), problem (G) must be posed in $\mathfrak{A}$.

Let u be a measurable and Lebesgue integrable function in A. The operator $\mathcal{E}(u)$ will be said to be definite on such a function, if there exists a function f, integrable on A, such that, for any function v belonging to $\mathcal{C}^*$, vanishing on Σ and such that $\mathcal{E}^*(v)$ is bounded on A, one has:

---

[7] Estimates of type (2.7.16) have been established by Picone. It is impossible, however, to infer (2.7.16) from the estimates of this author, since for their application one would have to prescribe the vanishing of u on a set containing Σ^*, but in general not coinciding with it.

$$\int_A (u\mathcal{E}^*(v) - vf)\, dx = 0. \tag{2.7.17}$$

In such a case we shall set $\mathcal{E}(u) = f$.[8]

It is an easy matter to see that $\mathcal{E}(u) = f$, $\mathcal{E}(u) = f'$ imply that f' and f coincide almost everywhere, i.e. the *equivalence* between f and f'. Therefore, if u is smooth enough, for instance belonging to $\mathcal{C}$, the operator $\mathcal{E}(u)$, just defined, is equivalent to $\mathcal{E}(u)$, intended in the classical sense. It follows that the generalized definition of $\mathcal{E}(u)$, given above, may actually be considered as an extension of such an operator.

Let $\mathfrak{U}$ be the class of the measurable functions on A, admitting an $\mathcal{E}(u)$ in the generalized sense just introduced. We shall consider problems (2.7.10)–(2.7.11) within such a class. To simplify the treatment, we shall assume, essentially without loss of generality, $\varphi \equiv 0$ on Σ^*.

Let us remark, anyway, that considerations, analogous to those we will come to discuss in a moment, may be repeated even letting φ be arbitrary and treating directly the general problem.

For a function of $\mathcal{C}$, the boundary condition

$$u = 0 \qquad \text{on } \Sigma^* \tag{2.7.11$_0$}$$

is equivalent, putting $\mathcal{E}(u) = f$, to the fact that (2.7.17) holds for any v belonging to the class V of the functions satisfying the following conditions:

(a) v belongs to $\mathcal{C}^*$, with $\mathcal{E}^*(v)$ bounded in A;

(b) v vanishes on $\Sigma^{(1)} \cup \Sigma^{(3)}$.

Now we shall say that $u \in \mathfrak{U}$ *satisfies* (2.7.11$_0$) *if* (2.7.17) *is satisfied for any* $v \in$ V.

We are, thus, led to the following generalized formulation of the boundary problem (2.7.10), (2.7.11$_0$).

(D) *Given the function* $f(x)$, *integrable on* A, *look for a function* $u(x)$, *integrable on* A, *satisfying* (2.7.17) *for any* $v \in$ V.

Denoting by $\mathfrak{L}^p(A)$ $(p \geqslant 1)$ the Banach space of the real functions, whose p-th power is integrable on A, let $f \in \mathfrak{L}^p(A)$ and assume that there exists a solution of problem (D) belonging to $\mathfrak{L}^q(A)$. Such a solution is defined to be a *weak* $[\mathfrak{L}^q, \mathfrak{L}^p]$ *solution* of problem (2.7.10), (2.7.11$_0$).

Consider also the Banach space $\Lambda(A \cup \Sigma)$, whose elements are completely additive functions $\varphi(B)$ real valued for any Borel set $B \subset A \cup \Sigma$. The norm in $\Lambda(A \cup \Sigma)$ is the following:

$$\|\varphi\| = v_\varphi(A \cup \Sigma),$$

where $v_\varphi(A \cup \Sigma)$ stands for the total variation of φ on $A \cup \Sigma$.

Problem (D) can be further generalized by considering, for any fixed $f \in \Lambda(A \cup \Sigma)$, the equations:

---

[8] Such extension of the operator $\mathcal{E}(u)$ corresponds to the Friedrichs *weak extension* of a differential operator.

$$\int_{A} u\mathcal{E}^*(v)\,dx - \int_{A\cup\Sigma} v\,df = 0,$$

with $v \in V$. Or else, looking also for the unknown u in $\Lambda(A \cup \Sigma)$, by considering the following other ones:

$$\int_{A\cup\Sigma} \mathcal{E}^*(v)\,du - \int_{A\cup\Sigma} v\,df = 0.$$

It is, then, clear what meaning is to be assigned to the term a *weak* $[\mathfrak{L}^q, \Lambda]$, or $[\Lambda, \mathfrak{L}^p]$, or $[\Lambda, \Lambda]$ *solution* of problem (2.7.10), $(2.7.11_0)$.

The following theorems are an immediate consequence of the existence principle of Section 2.

2.IX. *Let p and q be two real numbers greater than* 1. *A weak* $[\mathfrak{L}^q, \mathfrak{L}^p]$ *solution of problem* (2.7.10), $(2.7.11_0)$ *exists if, and only if, there exists, for any* $f \in \mathfrak{L}^p(A)$ *a positive constant* K *such that*

$$\left(\int_A |v|^{p/(p-1)}\,dx\right)^{(p-1)/p} \leqslant K\left(\int_A |\mathcal{E}^*(v)^{q/(q-1)}\,dx\right)^{(q-1)/q}, \tag{2.7.18}$$

for any $v \in V$.

For $p = \infty$ $(q = \infty)$, in which case $\mathfrak{L}^\infty(A)$ is the Banach space of pseudo-bounded functions in A, on the left-hand side (right-hand side) of (2.7.18) $\frac{p}{p-1}$ and $\frac{p-1}{p}$ $\left(\frac{q}{q-1}\text{ and }\frac{q-1}{q}\right)$ must be replaced by 1.

2.X. *A necessary and sufficient condition for the existence of a weak* $[\mathfrak{L}^q, \Lambda]$ *solution of problem* (2.7.10), $(2.7.11_0)$, *no matter how f is assigned in* $\Lambda(A \cup \Sigma)$, *is the existence of a positive constant* K *such that*

$$\max_{A\cup\Sigma} |v| \leqslant K\left(\int_A |\mathcal{E}^*(v)|^{q/(q-1)}\,dx\right)^{(q-1)/q},$$

for any $v \in V$.

Analogous theorems hold for the weak $[\Lambda, \mathfrak{L}^p]$ and $[\Lambda, \Lambda]$ solutions.

These theorems clearly show the whole importance of the estimates established by Theorem 2.VII and by (2.7.16). For, on applying them to the operator $\mathcal{E}^*$, sufficient conditions are obtained for the existence of weak solutions of problem (2.7.10), $(2.7.11_0)$.

For the case of the totally elliptic equations, the following theorem shows that the existence of a weak solution is equivalent to the existence of a solution in the classical sense.

2.XI. *If* $\mathcal{E}(u)$ *is totally elliptic in* $A \cup \Sigma$, *c being Hölder continuous in* $A \cup \Sigma$ *and non-positive there, if* Σ *is of class* $C^{1,h}$ *at any point and if f is Hölder continuous in* $A \cup \Sigma$, *any weak* $[\mathfrak{L}^q, \mathfrak{L}^p]$ *solution of problem* (2.7.10), $(2.7.11_0)$ *(Dirichlet problem) coincides almost everywhere with a function of class* 1 *in* $A \cup \Sigma$ *and of class* 2 *in* A, *satisfying* (2.7.1) *and vanishing on* Σ.

An analogous theorem holds for the classical parabolic equations such as the heat equation.

Coming back to the general case, as far as the uniqueness of a weak solution is concerned, the following theorem holds.

2.XII. *The weak* $[\mathfrak{L}^q, \mathfrak{L}^p]$ $([\mathfrak{L}^q, \Lambda])$ *solution of problem* (2.7.10), $(2.7.11_0)$ *is unique when, and only when, the range* $\mathcal{E}^*(V)$ *of the linear transformation* $\mathcal{E}^*(v)$, *considered for* $v \in V$, *is dense in the space* $\mathfrak{L}^{q/(q-1)}(A)$.

For the case of the weak $[\Lambda, \mathfrak{L}^p]$ or $[\Lambda, \Lambda]$ solutions, uniqueness is ensured when, and only when, $\mathcal{E}^*(V)$ is dense in the Banach space $\Gamma(A \cup \Sigma)$ of the continuous functions on $A \cup \Sigma$.

If $\mathcal{E}(u)$ is totally elliptic in $A \cup \Sigma$, it happens, as implied by elliptic equation theory, that $\mathcal{E}^*(V)$ is dense in $\mathfrak{L}^s(A)$ $(s \geq 1)$ or in $\Gamma(A \cup \Sigma)$, if the uniqueness theorem holds in the class $\mathcal{C}$ for problem (2.7.10), $(2.7.11_0)$, which, in this particular case, reduce to the classical Dirichlet problem.

Such a circumstance occurs also for the parabolic equations of the classical type (heat equations), which are usually considered in applications. In other words the uniqueness theorem in $\mathcal{C}$, for the corresponding problem (2.7.10), $(2.7.11_0)$, implies that $\mathcal{E}^*(V)$ is dense in $\mathfrak{L}^s(A)$ and in $\Gamma(A \cup \Sigma)$.

The memoirs [155], [156] of the present author have been followed by a number of papers of several mathematicians: Il'in, Lieberstein, Oleinik, Phillips, Sarason, Nirenberg, Kohn, Radkevitch, Dufner. The principal aim of those papers was to study under what conditions the solution of problem (D) is a solution of (G) in the classical sense and under what conditions the circumstance enunciated in Theorem 2.XII is verified. For an exhaustive source of reference on this subject, we refer to the recent monograph of Oleinik and Radkevitch [158], which has been translated into English [193] and Italian [194].

Since in the present survey we want to consider only the quantitative aspects of the problem, we shall report only a numerical analysis investigation, due to Lieberstein [159], [160], which was proposed to that author by the writer.

It follows from the theory recalled above that a solution of the equation $\mathcal{E}(u) = f$ may be determined only by f, without prescribing boundary conditions on u in any part of Σ. This happens when $\Sigma \equiv \Sigma^{(1)}$.

We have already produced some examples in this direction. Suppose now $c < 0$ in $\bar{A}$, let $\Sigma \equiv \Sigma^{(3)}$ and let the domain $\bar{A}$ contain a simple, closed manifold Γ which is the boundary of a domain B and is, in addition, completely of type $\Sigma^{(1)}$.

If the boundary problem

$$\mathcal{E}(u) = 0 \qquad \text{in A}, \qquad u = g \qquad \text{on } \Sigma, \tag{2.7.19}$$

is considered, under suitable assumptions it admits a solution in the classical sense ([161], [158]). On the other hand, since the equation $\mathcal{E}(u) = 0$ admits in B only the solution $u \equiv 0$, the solution of problem (2.7.10) *must vanish identically in* B, *no matter how g has been prescribed.* This must happen even if $u \in C^\infty$ in A, as it may be verified in some particular cases.

Let us consider, for instance, equation (2.7.14), which will now be rewritten setting $x_1 = x$, $x_2 = y$, $c \equiv -1$, $f \equiv 0$:

$$\mathcal{E}(u) \equiv y^2 u_{xx} - 2xyu_{xy} + x^2 u_{yy} - 2xu_x - 2yu_y - u = 0. \tag{2.7.20}$$

Consider this equation in the domain A:

$$-1 < x < 1, \qquad -1 < y < 1.$$

Let us look for a solution of (2.7.20) coinciding with a given function on $\Sigma \equiv \partial A$. More precisely, denoting by $g(t)$ a function defined in $[-1, 1]$, let:

$$u(1, y) = g(y), u(x, 1) = g(x), \tag{2.7.21}$$

$$u(-1, y) = g(y), u(x, -1) = g(x).$$

Suppose that $g(t)$ is of class C^∞ in $[-1, 1]$ and that it is an even function of t: $g(t) = g(-t)$. Let it further be such that

$$g(0) = 0.$$

The following result holds [161]:

(a) *There exists a unique solution of* (2.7.20), (2.7.21) *continuous in* $\bar{A}$ *and belonging to* $C^\infty(A)$.

Since in the present case it may be assumed that Γ is the circle $x^2 + y^2 = 1$ and, hence, that B is the circular domain bounded by Γ, it follows from our theory that:

(b) *The above solution u vanishes identically in* B.

Let us make a brief mention of the way of obtaining the results (a) and (b), without making appeal, in the particular case under consideration, to general theories.

As is obvious, we can limit our considerations to the region Q: $x > 0, y > 0$ and show the existence of a solution of (2.7.20) in the part D of A–B contained in Q (*see* Fig. 2.7), which has to be continuous in $\bar{D} = \overline{(A - B) \cap Q}$, to belong to $C^\infty(D)$ and to satisfy the boundary conditions: $u(1, y) = g(y)$, $u(x, 1) = g(x)$ $(0 \leqslant x \leqslant 1, 0 \leqslant y \leqslant 1)$. Furthermore, this function u has to vanish together with all its partial derivatives on the circular arc Γ_0: $x^2 + y^2 = 1, x > 0, y > 0$, so that it may be continued in the domain C: $x^2 + y^2 < 1, x > 0, y > 0$, into the identically zero function, thus giving rise to a solution of (2.7.20) of class C^∞ in Q.

Performing the change of variables $x = \rho \cos\theta$, $y = \rho \sin\theta$, the domains C and D and the curves Γ_0, S_1, S_2 (S_1 = segment $x = 1, 0 \leqslant y \leqslant 1$; S_2 = segment $y = 1, 0 \leqslant x \leqslant 1$) are

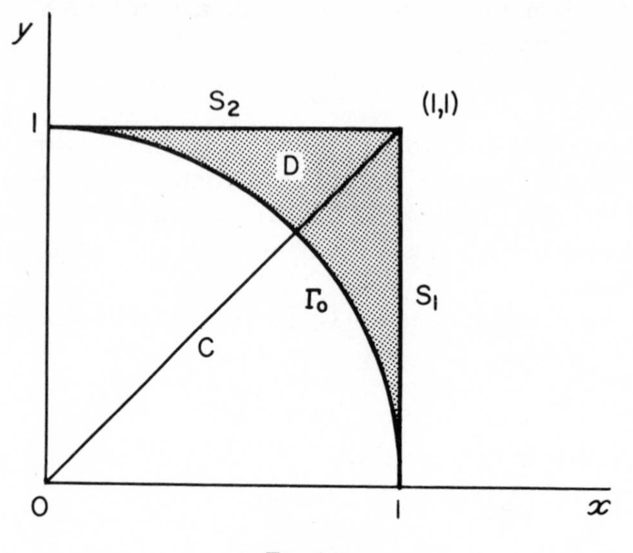

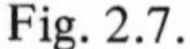
Fig. 2.7.

mapped into $D', C', \Gamma_0', S_1', S_2'$, as shown in Fig.2.8.

Equation (2.7.20) becomes:

$$\frac{\partial^2 u}{\partial \theta^2} - \rho \frac{\partial u}{\partial \rho} - u = 0 \tag{2.7.22}$$

and the boundary conditions become:

$$u = g(\rho \sin \theta) \text{ on } S_1', \text{ i.e. for } \rho = (\cos \theta)^{-1}, \; \left(0 \leqslant \theta \leqslant \tfrac{\pi}{4}\right) \tag{2.7.23}$$

$$u = g(\rho \cos \theta) \text{ on } S_2', \text{ i.e. for } \rho = (\sin \theta)^{-1}, \; \left(\tfrac{\pi}{4} \leqslant \theta \leqslant \tfrac{\pi}{2}\right).$$

(2.7.22) is a parabolic equation of the heat type and, furthermore, it may be reduced to this last type first through the change of variable $\rho = e^t$ and then through the substitution $v = e^t u$ for the unknown function. By (2.7.22), using the classical results on the heat equation and remembering that $g(0) = 0$, one may prove the existence (and the uniqueness) of a solution u belonging to $C^0(\bar{D}') \cap C^\infty(D')$, satisfying (2.7.23) and such that u vanishes together with its partial derivatives of all orders on Γ_0'.

Coming back to problem (2.7.20), (2.7.21), in view of the peculiar character of the phenomenon represented by the two results (a) and (b), the writer and Lieberstein decided during a common stay at the Mathematics Research Center of the University of Wisconsin, to verify the results arising from (a) and (b), through some numerical experiments. This was done in order to convince some mathematicians who were unwilling to follow the theoretical proofs of (a) and (b). The numerical experiments were undertaken by Lieberstein, who approximated the unknown function by means of polynomials. For $g(t)$ the function $g(t) \equiv t^2$ was chosen. It is to be remarked that, in view of the results to be achieved, the polynomials are not the best functions to take in order to approximate a function u which, for instance, must vanish on the diagonal of the square Q for $x^2 + y^2 < 1$, and which has later to rise very rapidly up to the value 1, reached by u at the vertex (1, 1).

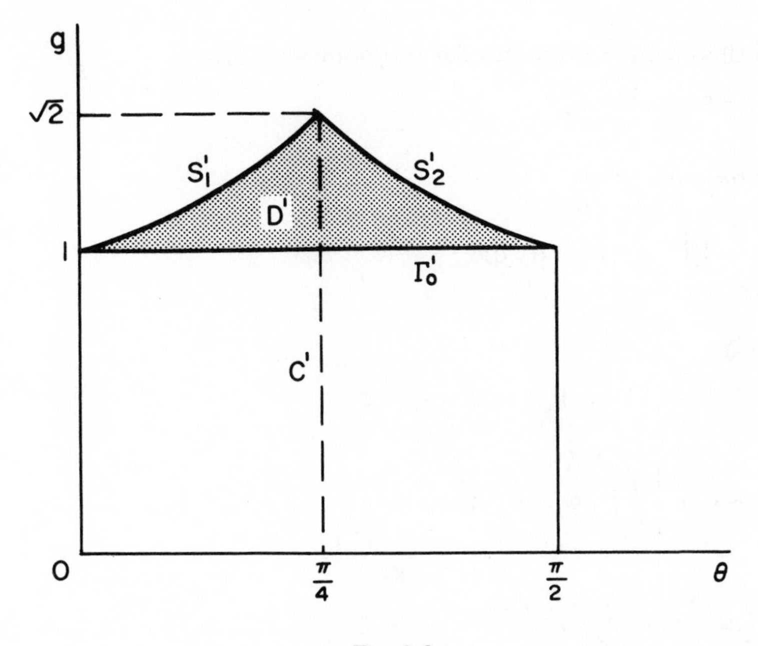

Fig. 2.8.

However, a different and more suitable choice of the approximating functions would make less credible the numerical experiment perfomed.

The approximating function was chosen in the following way:

$$u_n = \sum_{k=1}^{n} c_k v_k,$$

the v_k's being the functions in the following table:

Table 1

$v_1 = x^2 + y^2$	$v_9 = x^{10} + y^{10}$	$v_{17} = x^{12}y^2 + x^2y^{12}$	$v_{24} = x^8y^8$
$v_2 = x^4 + y^4$	$v_{10} = x^8y^2 + x^2y^8$	$v_{18} = x^{10}y^4 + x^4y^{10}$	$v_{25} = x^{18}y^{18}$
$v_3 = x^2y^2$	$v_{11} = x^6y^4 + x^4y^6$	$v_{19} = x^8y^6 + x^6y^8$	$v_{26} = x^{16}y^2 + x^2y^{16}$
$v_4 = x^6 + y^6$	$v_{12} = x^{12} + y^{12}$	$v_{20} = x^{16} + y^{16}$	$v_{27} = x^{14}y^4 + x^4y^{14}$
$v_5 = x^4y^2 + x^2y^4$	$v_{13} = x^{10}y^2 + x^2y^{10}$	$v_{21} = x^{14}y^2 + x^2y^{14}$	$v_{28} = x^{12}y^6 + x^6y^{12}$
$v_6 = x^8 + x^8$	$v_{14} = x^8y^4 + x^4y^8$	$v_{22} = x^{12}y^4 + x^4y^{12}$	$v_{29} = x^{10}y^8 + x^8y^{10}$
$v_7 = x^6y^2 + x^2y^6$	$v_{15} = x^6y^6$	$v_{23} = x^{10}y^6 + x^6y^{10}$	$v_{30} = x^{20} + y^{20}$
$v_8 = x^4y^4$	$v_{16} = x^{14}y^{14}$		

The constants c_k were obtained through a least square procedure:

$$\int_{\Sigma}\left[g-\sum_{k=1}^{n} c_k v_k\right]^2 ds + \iint_{A}\left[\sum_{k=1}^{n} c_k\, \mathcal{E}(v_k)\right]^2 dx\, dy = \text{minimum}. \tag{2.7.24}$$

The general theory gave estimates for the pointwise error

$$P(u-u_n) = \max_{\bar{A}} |u-u_n|$$

and for the global one

$$I(u-u_n) = \iint_{A} |u-u_n|^2\, dx\, dy.$$

More precisely

$$P(u-u_n) \leqslant \max_{\bar{A}} |\mathcal{E}(u_n)| + \max_{\Sigma} |u_n - g|$$

$$I(u-u_n) \leqslant 0.5 \left\{\iint_{D} |\mathcal{E}(u_n)|^2\, dx\, dy + \int_{S_1 \cup S_2} |g-u_n|^2\, ds + \int_{\Gamma_0} |u_n|^2\, ds\right\} + \pi \max_{\bar{\Gamma}_0} |u_n|^2 .$$

The following tables show the various approximations obtained. At the 30th approximation Lieberstein noticed that, as often occurs with least squares methods, the approximation was worsening near the angles. In order to avoid such an occurrence, he used a particular technique, which he called 'boundary extension', consisting of replacing in the left-hand side of (2.7.24) the integral extended to Σ by another one extended to a wider set.

The graphs of $u_n(x, y)$ are also shown, corresponding to some 'sections' of the surface $z = u_n(x, y)$. Finally Table 8 shows the numerical results obtained by considering the operator:

$$\mathcal{E}(u) + \delta(x, y)\, \Delta_2 u$$

($\delta(x, y)$ function of class C^2, with support in B). The two results (a) and (b) hold also for this operator.

Table 2

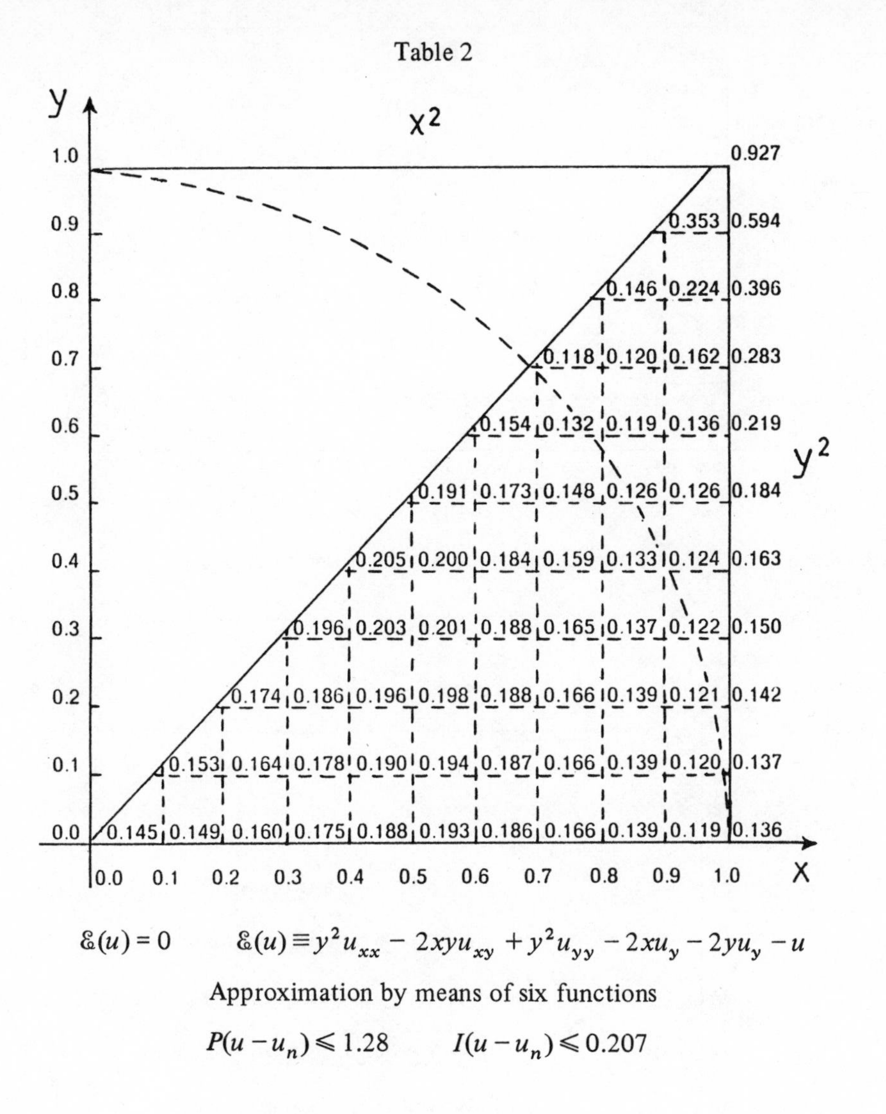

$\mathcal{E}(u) = 0 \qquad \mathcal{E}(u) \equiv y^2 u_{xx} - 2xyu_{xy} + y^2 u_{yy} - 2xu_y - 2yu_y - u$

Approximation by means of six functions

$P(u - u_n) \leqslant 1.28 \qquad I(u - u_n) \leqslant 0.207$

Table 3

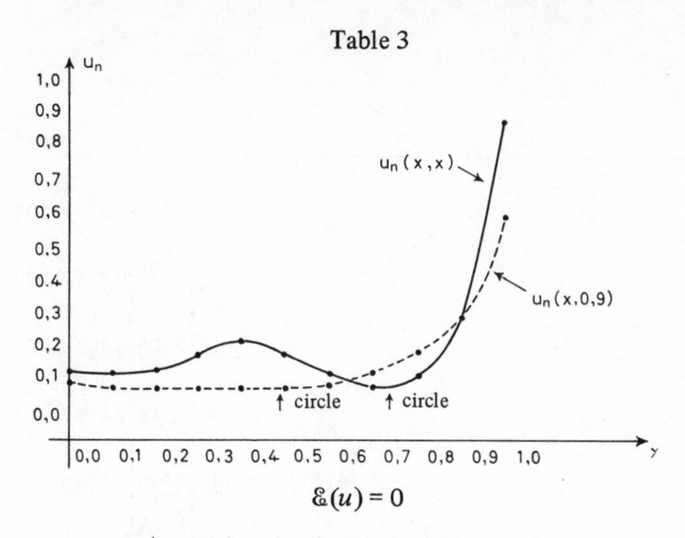

$\mathcal{E}(u) = 0$

Approximation by means of six functions

Table 4

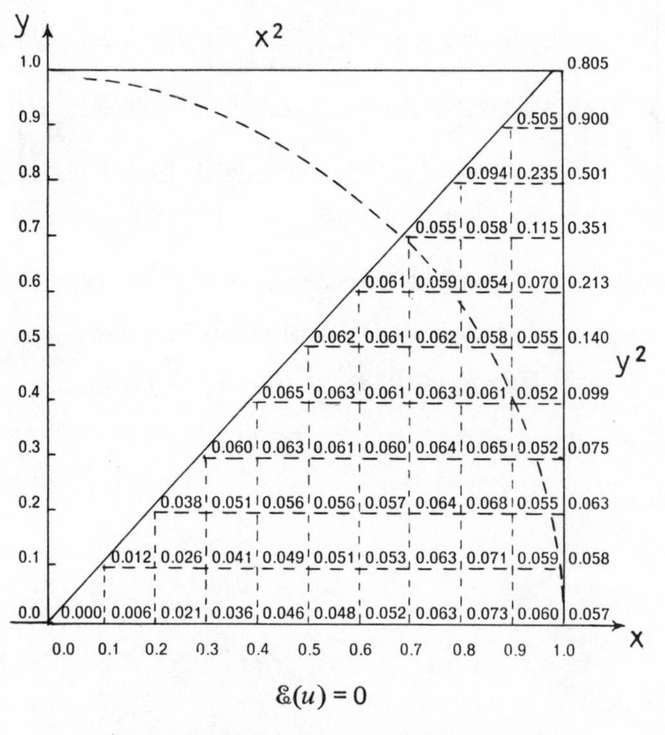

$\mathcal{E}(u) = 0$

Approximation by means of thirty functions

$P(u-u_n) \leqslant 1.18 \qquad I(u-u_n) \leqslant 0.033$

Table 5

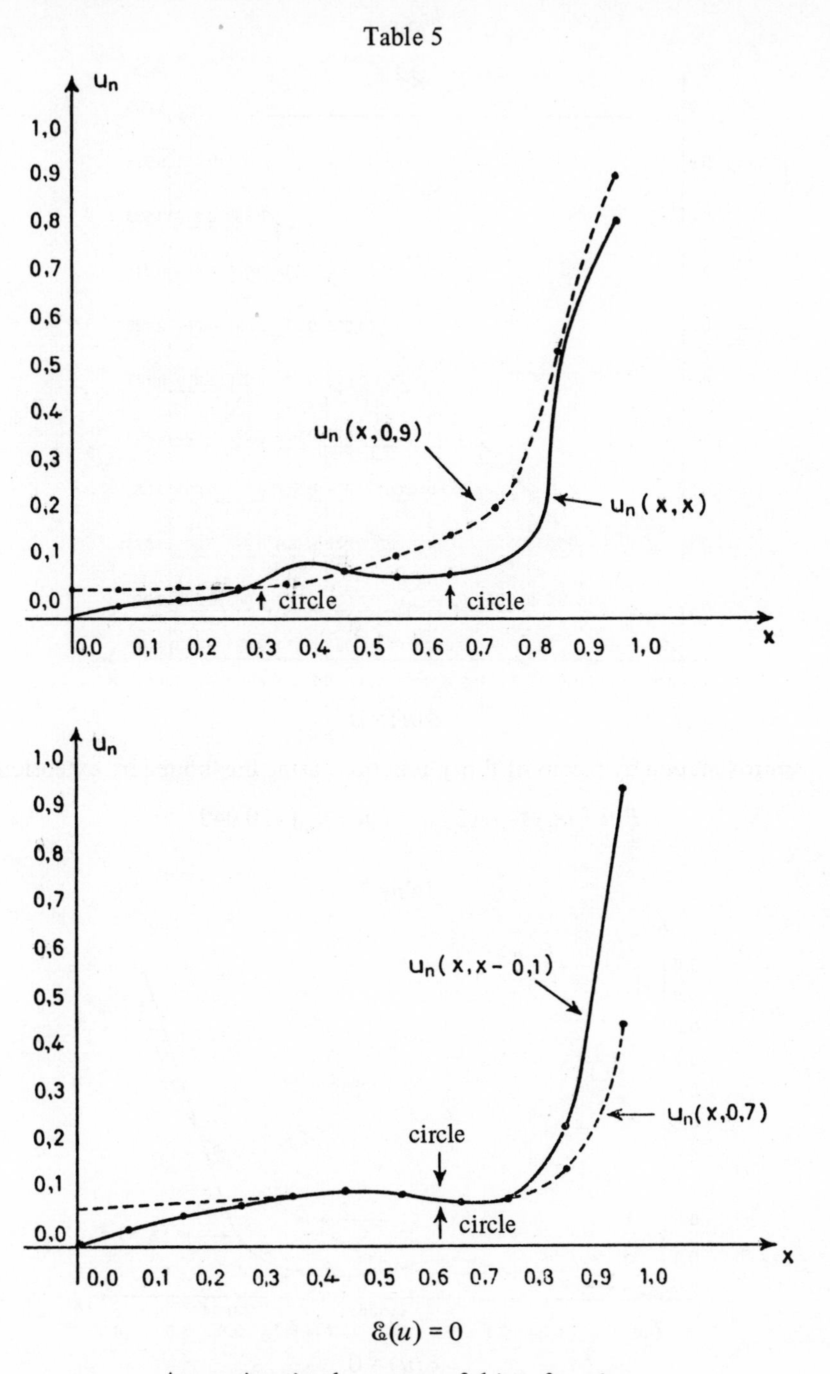

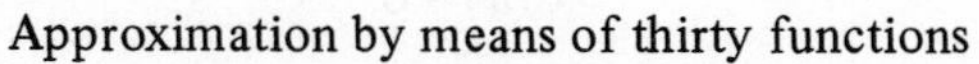

$$\mathfrak{E}(u) = 0$$

Approximation by means of thirty functions

Table 6

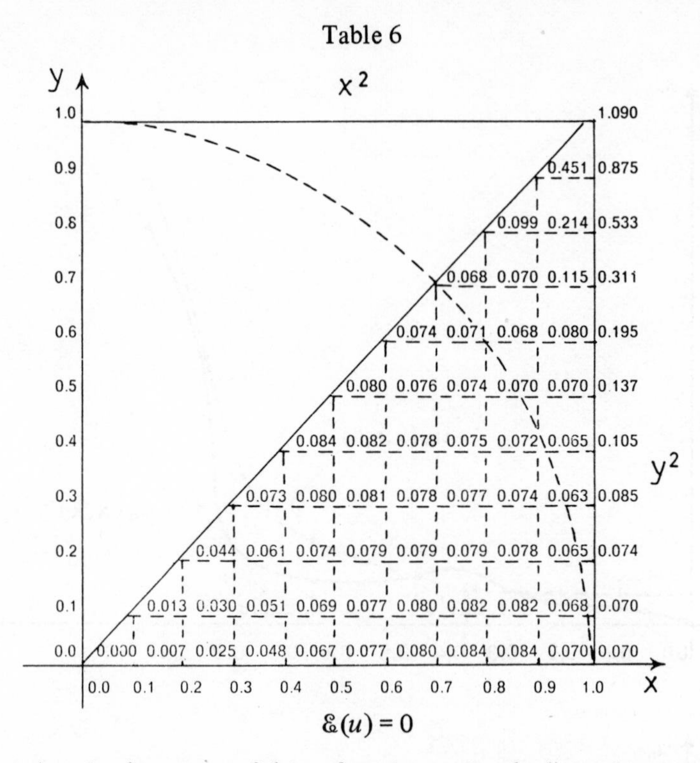

$\mathscr{E}(u) = 0$

Approximation by means of thirty functions using the 'boundary extension'

$$P(u - u_n) \leqslant 1.12 \qquad I(u - u_n) \leqslant 0.042$$

Table 7

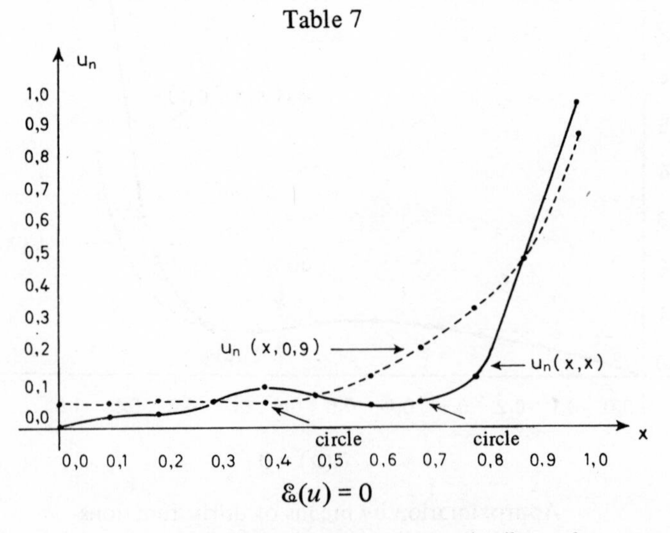

$\mathscr{E}(u) = 0$

Approximation by means of thirty functions using the 'boundary extension'

Table 8

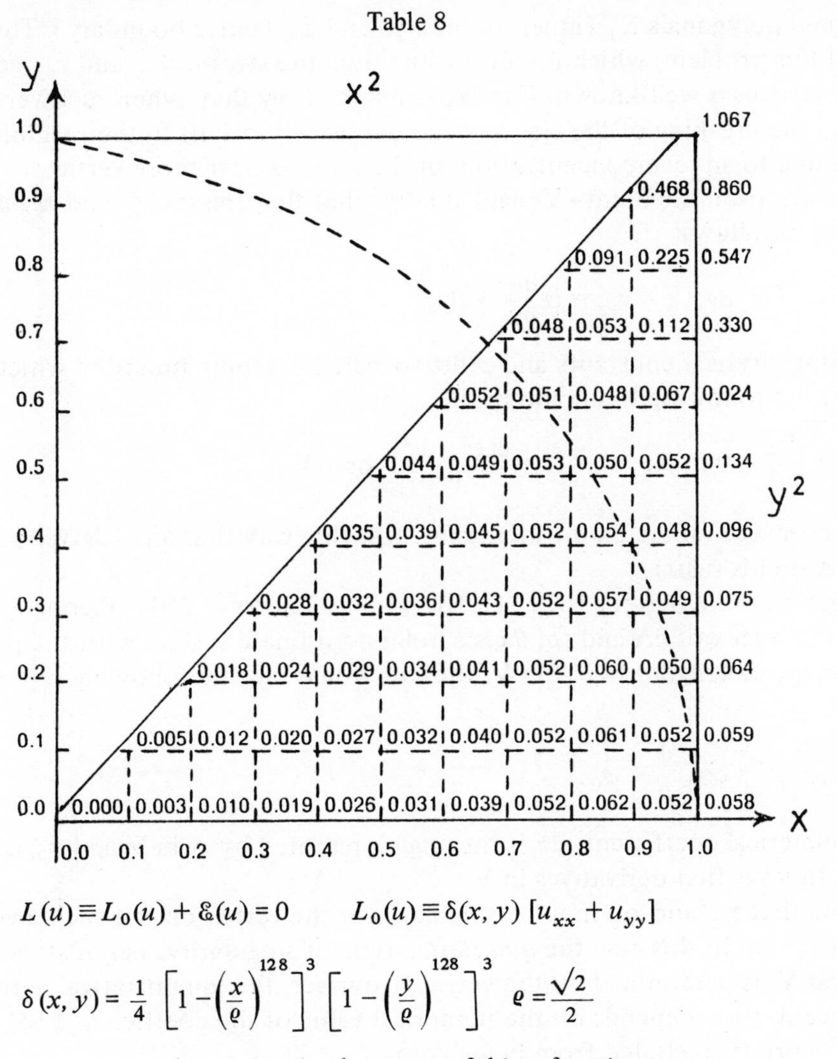

$$L(u) \equiv L_0(u) + \mathcal{E}(u) = 0 \qquad L_0(u) \equiv \delta(x, y)\,[u_{xx} + u_{yy}]$$

$$\delta(x, y) = \frac{1}{4}\left[1 - \left(\frac{x}{\varrho}\right)^{128}\right]^3 \left[1 - \left(\frac{y}{\varrho}\right)^{128}\right]^3 \qquad \varrho = \frac{\sqrt{2}}{2}$$

Approximation by means of thirty functions

8. Estimates of Physical Quantities: The Stress Concentration Coefficient in an Incoming Vertex of a Prism Under Torsion

In this Section, as well as in the following one, some results will be considered concerning the calculation of some physical quantities, whose values are theoretically obtained by solving partial differential equation problems. They are not included within those physical quantities (eigenfrequencies of vibrating bodies), whose computation is provided by the methods reviewed in Part I.

Let A be the plane cross-section of a hollow slender cylinder under torsion, to which the Saint–Venant theory can be applied. Let A have a boundary ∂A, formed by two simple closed polygonals Σ_1 (inner boundary) and Σ_2 (outer boundary). The technical interest of this problem, which involves estimating the stresses τ_{xz} and τ_{yz} near the boundary vertices is well known. For experiments show that, when such vertices are 'incoming', the breaking of the elastic structure usually occurs in their vicinity. These breaks are due to an 'overconcentration' of the stresses near those vertices.

It is known, from the Saint–Venant theory, that the stresses τ_{xz} and τ_{yz} may be represented as follows:

$$\tau_{xz} = \alpha\frac{\partial\varphi}{\partial x} - \beta y, \qquad \tau_{yz} = \alpha\frac{\partial\varphi}{\partial y} + \beta x,$$

α and β being physical constants and φ the so-called 'warping function' which is the solution of the following Neumann problem in A:

$$\frac{\partial^2\varphi}{\partial x^2} + \frac{\partial^2\varphi}{\partial y^2} = 0 \text{ in A}, \qquad \frac{\partial\varphi}{\partial\nu} = \frac{\partial(x^2 + y^2)}{\partial s} \text{ on } \partial\text{A}$$

(ν inward normal, s an arc on ∂A increasing in such a way that an observer moving along ∂A leaves A on his right).

Investigations by Koetter [162], Trefftz [163], Tricomi [164], Picone [165] prove that, if V is a vertex of ∂A and (ρ, θ) is a polar coördinate system with the pole in V and the polar axis tangent to ∂A in V, the function φ has the following representation near V:

$$\varphi = q[\varphi]\,\rho^{1/\mu}\cos\frac{\theta}{\mu} + \varphi_0;$$

$q[\varphi]$ is a numerical coefficient, $\mu\pi$ is the angle presented by A in V and φ_0 is a function having continuous first derivatives in V.

It follows that τ_{xz} and τ_{yz} may be singular near the vertex only if such a vertex is incoming ($\mu > 1$). In this case the *qualitative* type of singularity, i.e. of stress 'concentration' near V, is determined by the vertex. However, the 'quantitative' nature of such a concentration depends on the numerical value of the coefficient $q[\varphi]$, which cannot be 'a priori' excluded from being zero.

Paper [166] is devoted to the problem concerning the construction of two sequences $\{q'_k[\varphi]\}$ and $\{q''_k[\varphi]\}$, converging to $q[\varphi]$ and such that

$$q'_k[\varphi] < q[\varphi] < q''_k[\varphi].$$

Let u and v be the solutions of the following Dirichlet problems:

$$\Delta_2 u = 0 \text{ in A}, \qquad u = x^2 + y^2 \text{ on } \partial\text{A};$$

$$\Delta_2 v = 0 \text{ in A}, \qquad v = 0 \text{ on } \Sigma_1, \qquad v = 1 \text{ on } \Sigma_2.$$

Denoting by C any closed curve, contained in A and enclosing Σ_1, it turns out that

$$\int_C \frac{\partial v}{\partial\nu}\, ds = 0.$$

Set

$$\alpha = \int_C \frac{\partial u}{\partial \nu}\, ds \left(\int_C \frac{\partial v}{\partial \nu}\, ds \right)^{-1}$$

The function $\psi = u - \alpha v$ is the harmonic conjugate of φ. In the neighbourhood of V one has:

$$u = q[u]\rho^{1/\mu} \sin \frac{\theta}{\mu} + u_0, \qquad v = q[v]\rho^{1/\mu} \sin \frac{\theta}{\mu} + v_0,$$

$q[u]$ and $q[v]$ being numerical coefficients and u_0 and v_0 functions with continuous first derivatives in V. One has (assuming $V \subset \Sigma_1$):

$$q[u] = \frac{2}{\mu\pi}\, \rho^{-1/\mu} \int_0^{\mu\pi} u(\rho, \theta) \sin \frac{\theta}{\mu}\, d\theta + F(\rho),$$

$$q[v] = \frac{2}{\mu\pi}\, \rho^{-1/\mu} \int_0^{\mu\pi} v(\rho, \theta) \sin \frac{\theta}{\mu}\, d\theta, \qquad q[\varphi] = q[\psi] = q[u] - \alpha q[v],$$

$F(\rho)$ is a known function of ρ. Obviously $q[v] > 0$.

The two sequences $\{q'_k[\varphi]\}$ and $\{q''_k[\varphi]\}$ can be obtained as soon as the following problems are solved:

(1) construct $\{q'_k[u]\}$ and $\{q''_k[u]\}$ convergent to $q[u]$ and such that:

$q'_k[u] \leqslant q[u] \leqslant q''_k[u]$;

(2) construct $\{q'_k[v]\}$ and $\{q''_k[v]\}$ convergent to $q[v]$ and such that:

$q'_k[v] \leqslant q[v] \leqslant q''_k[v]$;

(3) construct $\{\alpha'_k\}$ and $\{\alpha''_k\}$, convergent to α and such that:

$\alpha'_k \leqslant \alpha \leqslant \alpha''_k$.

Consider the system of functions formed by $\omega_{-1} = \log |z|$ (the point $z = 0$ is considered inside the domain whose boundary is Σ_1) and $\omega_k (k = 0, 1, 2, \ldots)$, where the ω_k are obtained by arranging into one sequence the functions

$$\Re\, z^h, \qquad \Im\, z^h \qquad (h = -2, -1, 0, 1, 2, \ldots).$$

Let

$$u_n = \sum_{k=-1}^{n} c_k^{(n)} \omega_k, \qquad v_n = \sum_{k=-1}^{n} \gamma_k^{(n)} \omega_k$$

where $c_k^{(n)}$ and $\gamma_k^{(n)}$ have been obtained by imposing the condition

$$\int_{\partial A} \left(|u - u_n|^2 + \left| \frac{\partial u}{\partial s} - \frac{\partial u_n}{\partial s} \right|^2 \right) ds = \text{minimum},$$

$$\int_{\partial A} \left(|v - v_n|^2 + \left| \frac{\partial v}{\partial s} - \frac{\partial v_n}{\partial s} \right|^2 \right) ds = \text{minimum}.$$

A suitable completeness theorem makes sure that u_n and v_n converge towards u and v, respectively, uniformly in $\bar{A}$.

Let

$$m_n = \max_{\partial A} |u - u_n|, \qquad \mu_n = \max_{\partial A} |v - v_n|.$$

Set

$$u_n^{(1)} = u_n - m_n, \qquad u_n^{(2)} = u_n + m_n,$$

$$v_n^{(1)} = u_n - \mu_n, \qquad v_n^{(2)} = u_n + \mu_n.$$

One has:

$$u_n^{(1)} \leqslant u \leqslant u_n^{(2)}, \qquad v_n^{(1)} \leqslant v \leqslant v_n^{(2)},$$

$u_n^{(1)}$ and $u_n^{(2)}$ converging to u uniformly in $\bar{A}$, $v_n^{(1)}$ and $v_n^{(2)}$ converging to v, uniformly in $\bar{A}$.

The sequences, which solve problems (1) and (2), are obtained by setting:

$$q_k'[u] = \frac{2}{\mu\pi} \rho^{-1/\mu} \int_0^{\mu\pi} u_k^{(1)}(\rho, \theta) \sin \frac{\theta}{\mu}\, d\theta + F(\rho),$$

$$q_k''[u] = \frac{2}{\mu\pi} \rho^{-1/\mu} \int_0^{\mu\pi} u_k^{(2)}(\rho, \theta) \sin \frac{\theta}{\mu}\, d\theta + F(\rho),$$

$$q_k'[v] = \frac{2}{\mu\pi} \rho^{-1/\mu} \int_0^{\mu\pi} v_k^{(1)}(\rho, \theta) \sin \frac{\theta}{\mu}\, d\theta,$$

$$q_k''[v] = \frac{2}{\mu\pi} \rho^{-1/\mu} \int_0^{\mu\pi} v_k^{(2)}(\rho, \theta) \sin \frac{\theta}{\mu}\, d\theta;$$

ρ is arbitrarily fixed, but in such a way that the points (r, θ) for which $0 < r \leqslant \rho, 0 < \theta < \mu\pi$, are contained in A.

To construct $\{\alpha_k'\}$ and $\{\alpha_k''\}$, set:

$$U_n = b_{-1}^{(n)} v + \sum_{k=0}^{n} b_k^{(n)} \omega_k$$

and determine $b_{-1}^{(n)}$ and $b_k^{(n)}$ by imposing the condition:

$$\int_{\partial A} \left(|u - U_n|^2 + \left| \frac{\partial u}{\partial s} - \frac{\partial U_n}{\partial s} \right|^2 \right) ds = \text{minimum}.$$

A completeness theorem, analogous to the above-mentioned one, ensures the uniform convergence in $\bar{A}$ of U_n to u.

Assuming:

$$\alpha_k' = b_{-1}^{(k)} - \max_{\Sigma_2} (U_n - u) + \min_{\Sigma_1} (U_n - u)$$

$$\alpha_k'' = b_{-1}^{(k)} + \max_{\Sigma_1} (U_n - u) - \min_{\Sigma_1} (U_n - u)$$

the two sequences solving problem (3) are obtained.

The numerical applications made at the Istituto di Calcolo of the Italian Research Council (by means of desk calculators) lead to the following result, when Σ_1 and Σ_2 are two homothetic squares of half side 1 and 2:

$$2.13 < -q[\varphi] < 2.15.$$

Subsequently the method has been applied to other interesting particular cases by De Schwarz [167]. If Σ_1 and Σ_2 are two regular homothetic polygons, whose side number m is large enough and whose ratio R between the sides satisfies a suitable condition, De Schwarz has shown that some analytical estimates, depending on R and m, can be given for $q[\varphi]$.

9. Estimates of Physical Quantities: The Electrostatic Capacity of a Conductor

Let A be a bounded domain of the three-dimensional cartesian space x_1, x_2, x_3. Let the boundary of A be the simple closed surface Σ. Denote by E the set of all points lying outside A.

Consider the *exterior Dirichlet problem*

$$\frac{\partial^2 u}{\partial x_1^2} + \frac{\partial^2 u}{\partial x_2^2} + \frac{\partial^2 u}{\partial x_3^2} = 0 \qquad \text{in E} \tag{2.9.1}$$

$$u = 1 \qquad \text{on } \Sigma \tag{2.9.2}$$

$$u(\infty) = 0. \tag{2.9.3}$$

If ν denotes the outward normal to Σ, the *electrostatic capacity* C of Σ is defined, as is known, in the following way:

$$C = -\frac{1}{4\pi} \int_{\Sigma} \frac{\partial u}{\partial \nu} \, d\sigma.$$

It is well known, at least when Σ is smooth enough, that the function

$$\delta(x) = -\frac{1}{4\pi} \frac{\partial u}{\partial \nu}$$

expresses the density of the simple layer potential

$$\int_{\Sigma} \frac{\delta(x)}{|x-y|} \, d\sigma_x,$$

which represents the constant potential equal to 1 in A. Hence the capacity may also be defined as the ratio between the total charge on Σ and the constant expressing the value of the potential in A generated by that charge. For a sphere of radius R the value of C is immediately computed and one has: C = R.

The constant C occurs not only in electrostatics, but also in several other fields of application, such as, for instance, dynamics of fluids, electron optics, plasmadynamics, antenna theory, and so on. Therefore the actual computation of C is a problem of remarkable interest. Such a problem may be posed in two different ways: one corresponding to a predominantly engineering standpoint, which only requires the construction of some real number approximating C, without concern about bounds on the approximation error; the other one, on the contrary, is of a more specific mathematical nature and requires a majorization of such an error, or, at least, some information about the nature of the approximation obtained, i.e. whether one has to deal with an upper or a lower bound.

It is easily understood that the second formulation of the problem, of a purely analytical character, offers an higher order of difficulty with respect to the first one.[9]

Such a mathematical formulation gave rise to valuable investigations, begun by the fundamental papers of Polya [169] and Polya and Szegö [170], [171]. These authors, assuming that Σ is the cube of side 1, obtain for the capacity of Σ, indicated now by C^*, the following bounds:

$$0.62033 < C^* < 0.71055.$$

Later on, McMahon [172] obtained the following rigorous lower bound:

$$0.639273 < C^*.$$

Gross [173], applying a method of Picone [123], obtained for C^* the following approximate value:

---

(9) The first point of view (the engineering approach) is followed by Greenspan [168] who is concerned with the computation of C for a cube of unit side. By using a finite difference method, he proposes for C the value 0.661. No estimate is given in [168] for the approximation error.

$$C^* \simeq 0.6464$$

with the following error estimate:

$$|C^* - 0.6464| < 0.032.$$

The method, employed by Gross in obtaining this estimate, is not a mathematically rigorous one, but his paper, valuable from the numerical point of view, contains some interesting remarks which were later exploited adroitly by Payne and Weinberger [174] and allowed these authors to obtain rigorous bounds. Explicitly, they have obtained for C^*:

$$0.627 < C^* < 0.668.$$

In 1953 Daboni [175] published an interesting paper dealing with the computation of C^*, obtained through the following general theorem:

2.XIII. *Let $\alpha(B)$ be a completely additive set function, defined on the family $\{B\}_{\bar{A}}$ of all Borel sets contained in $\bar{A}$. Let α be such that the function*

$$v(y) = \int_A \frac{d_x \alpha}{|x - y|}$$

is continuous and positive in $\bar{E}$. Then the following bounds hold:

$$\frac{\alpha(\bar{A})}{\max_\Sigma v} \leqslant C \leqslant \frac{\alpha(\bar{A})}{\min_\Sigma v}. \tag{2.9.4}$$

If Σ is the cube of side 1, Daboni applied this theorem, taking as α a linear combination

$$\alpha = k_1 \alpha_1 + k_2 \alpha_2 + k_3 \alpha_3$$

of three measures $\alpha_1, \alpha_2, \alpha_3$; α_1 is the measure corresponding to a uniform charge distribution, with constant density equal to 1, on the faces of the cube Σ; α_2 corresponds to a uniform charge distribution, with density 1, on the edges of the cube, homothetic to Σ, of side 0.98; α_3 corresponds to eight unit point charges concentrated on the vertices of the above-mentioned cube of side 0.98.

Decomposing any face of Σ in 100 equal squares and laying down a unit charge on any vertex of the lattice thus obtained, if β is the measure obtained in this way, the constants k_1, k_2, k_3 are computed by Daboni imposing the condition

$$\int_\Sigma [v(x) - 1]^2 \, d\beta = \text{minimum},$$

where

$$v(x) = \sum_{i=1}^{3} k_i \int_{\Sigma} \frac{d_y \alpha_i}{|x - y|}.$$

One has then: $k_1 = 0.777125$, $k_2 = 0.0016036$, $k_3 = 0.0014295$.

Assume that Σ has like 'extrema' the two points (0, 0, 0) and (1, 1, 1), and denote by x, y, z the cartesian coordinates in X^3. Daboni obtains the following table (Table 9) of values of v which, for obvious symmetry reasons, can be considered only for $0 \leqslant x \leqslant \frac{1}{2}, 0 \leqslant y \leqslant x, z = 0$.

Table 9

0.5						1.0077
0.4					1.0057	1.0067
0.3				0.9998	1.0028	1.0039
0.2			0.9907	0.9955	0.9988	1.0000
0.1		0.9849	0.9880	0.9936	0.9975	0.9989
$y = 0.0$	1.0034	0.9924	0.9990	1.0066	1.0115	1.0131
$x =$	0.0	0.1	0.2	0.3	0.4	0.5

Such results led Daboni to assume for v the bounds:

$$0.98 < v < 1.0131$$

and, hence, by (2.9.4)

$$0.654 < C^* < 0.676.$$

Unfortunately, while the lower bound on v is correct and, hence, so also is the upper approximation on C^*, the bound $v < 1.0131$ does not hold, as was shown later by Sneider [176]. For, if the Daboni function v is computed (by means of an electronic computer) for $0 \leqslant x \leqslant 0.011$ and for $0 \leqslant y \leqslant x$, with step $q = 10^{-3}$, Table 10 is obtained:

Table 10

0.011												1.1198
0.010											1.1170	1.1184
0.009										1.1109	1.1139	1.1153
0.008									1.1018	1.1062	1.1092	1.1107
0.007								1.0905	1.0960	1.1003	1.1032	1.1047
0.006							1.0778	1.0840	1.0892	1.0932	1.0960	1.0975
0.005						1.0647	1.0710	1.0767	1.0816	1.0854	1.0881	1.0896
0.004					1.0513	1.0578	1.0637	1.0690	1.0735	1.0770	1.0796	1.0811
0.003				1.0387	1.0448	1.0507	1.0561	1.0610	1.0651	1.0684	1.0709	1.0724
0.002			1.0264	1.0323	1.0380	1.0434	1.0484	1.0528	1.0566	1.0597	1.0621	1.0636
0.001		1.0147	1.0204	1.0258	1.0311	1.0361	1.0406	1.0447	1.0482	1.0511	1.0533	1.0548
y = 0.000	1.0034	1.0089	1.0141	1.0192	1.0240	1.0286	1.0328	1.0366	1.0398	1.0425	1.0446	1.0461
x =	0.000	0.001	0.002	0.003	0.004	0.005	0.006	0.007	0.008	0.009	0.010	0.011

Since at the point $x = 0.011$, $y = 0.011$, v takes a value of not less than 1.119, the lower bound which can be obtained by (2.9.4) using v cannot be larger than 0.593.

It must be added that the remark by Sneider, concerning the error in the lower bound given by Daboni, does not diminish the value of the work of this author, who performed the numerical computations by means of a desk computer, about twenty-five years ago. In addition, it is a singular fact that the numerical flow was not discovered by De Schwarz, who at the Istituto di Calcolo of the Italian Research Council checked all numerical computations of the Daboni paper (*see* [175], page 466) before its publication, as well as by the several authors, who later examined Daboni's paper, some of them even basing their investigations upon it (Payne and Weinberger [174], Diaz [177], [178], [179], Protter and Weinberger [180], Parr [181]).

In 1961 Parr, a pupil of Diaz, studying deeply and further improving the methods of Polya and Szegö as well as others, achieved the following majorization of C^*:

$$C^* < 0.66755.$$

Prior to 1970, when the memoir [176] by Sneider was published, the best known bounds on C^* were, then, the following ones:

$$0.639273 \text{ (McMahon)} < C^* < 0.66755 \text{ (Parr)}.$$

In the memoir [176] several methods for computing C are considered. Some of them yield lower bounds, some others upper bounds and some others again approximate values of C, for which the approximation error is majorized.

As far as the unit cube is concerned, i.e. for $C = C^*$, as a result of the various procedures employed in [176], the best bounds obtained are:

$$0.6534 < C^* < 0.666002.$$

Both upper and lower bounds improve the formerly known results.

Sneider, before beginning the study of the analytical methods providing the rigorous computation of C for a large class of closed surfaces Σ made sure that the potential function u enjoys all those qualitative properties, required in the papers by the above-mentioned authors, as well as in [176]. For instance, in some papers mentioned above, use was made without further comment, of the Green identity

$$\int_\Sigma u \frac{\partial v}{\partial \nu} \, d\sigma = \int_\Sigma v \frac{\partial u}{\partial \nu} \, d\sigma,$$

where u is the potential function solution of problem (2.9.1), (2.9.2), (2.9.3) and v is a harmonic function in E, vanishing at infinity and of class 1 in the closure $\bar{E}$ of E. Now there is no doubt that, if Σ has no singularities and is smooth enough, for instance if Σ is of class $C^{1,h}$, u enjoys such smoothness properties that the Green formulae will hold.

Such a circumstance, however, is far from being evident as soon as Σ exhibits some singularities. Already for the cube case, u cannot be expected to belong to the class 1 in $\bar{E}$.

Thus, before studying the quantitative aspect of the problem, a detailed analysis was performed in order to achieve an existence theorem for the potential function u, which could establish all those qualitative properties of u, that are needed fully to justify the formal procedures (use of the Green formulae, continuity and integrability properties and so on) employed in the investigation.

This detailed analysis turned out to be a rather delicate one and its achievement required the precise formulation of well determined assumptions to be satisfied by Σ. These assumptions, although apparently complicated, allow such a generality to Σ that all cases usually met in applications are covered.

The characterization of the function class containing the solution of problem (2.9.1), (2.9.2), (2.9.3) is given by a theorem of remarkable theoretical interest.[10]

By way of example, it is proved among other results, that the 'density' $\delta(x) = -\frac{1}{4\pi}\frac{\partial u}{\partial \nu}$ belongs to $\mathfrak{L}^2(\Sigma)$, even when Σ exhibits vertices and edges, provided, of course, that these are not too 'bad'. To give a brief indication of the degree of generality allowed to A, it is enough to say that Sneider's results hold if $A \cup \Sigma$ is C^2-homeomorphic to the closure of a three-dimensional domain, bounded by a polyhedric surface.

Let us briefly mention here the approximation methods and the estimates for C considered by this author, reporting also the numerical results obtained for C^*.

Let $\varphi_0, \varphi_1, \ldots, \varphi_k, \ldots$ be a sequence of functions of class $C^1(\bar{E})$, harmonic in E and vanishing at infinity, complete in the space V(E) of the functions v, harmonic in E and such that

$$\int_E |\operatorname{grad} v|^2 dx < +\infty.$$

For instance, if the origin of the space is contained in A, it may be assumed that

$$\varphi_0 = \frac{1}{|x|}, \qquad \varphi_k(x) = \frac{P_k(x)}{|x|^{2\nu_k+1}} \qquad (k = 1, 2, \ldots), \tag{2.9.5}$$

$\{P_k(x)\}$ being a complete system of harmonic polynomials, homogeneous (of degree $\geqslant 1$) and ν_k the degree of P_k.

Setting, for $v, w \in V(E)$,

$$(v, w) = \int_E \operatorname{grad} v \operatorname{grad} w \, dx \tag{2.9.6}$$

[10] The procedures of Sneider have been applied by Castellani Rizzonelli [200] to the three-dimensional elasticity equations, obtaining for this case, although algorithmically more involved, results analogous to Sneider's.

and denoting by $G(w_1, \ldots, w_m)$ the Gram determinant, with respect to the scalar product (2.9.6), of the functions $w_1, \ldots, w_m$ belonging to V(E), Sneider proves that

$$4\pi \frac{G(\varphi_1, \ldots, \varphi_n)}{G(\varphi_0, \varphi_1, \ldots, \varphi_n)} \leqslant C. \tag{2.9.7}$$

The limit of the left-hand side as $n \to \infty$ is C, the convergence being monotonic increasing.

A geometric interpretation of this theorem in the function space V(E) is the following. Denoting by $V_0(E)$ the subspace of V(E) formed by all functions v, such that

$$\int_{\Sigma'} \frac{\partial v}{\partial \nu}\, d\sigma = 0,$$

Σ' being a surface homologous to Σ, contained in E. Denote by d the distance (with respect to the metric induced by (2.9.6)) between φ_0 and $V_0(E)$. One has:

$$C = 4\pi d^{-2}.$$

After having chosen, among the φ_k's given by (2.9.5), only those having all the symmetries of the cube centred in the origin, whose edges are parallel to the axes (choice mechanized through a computer [182]) Sneider obtains, applying (2.9.7), the following numerical lower bounds on C^*:

$n = 0$	$0.601547 < C^*$
$n = 1$	$0.622948 < C^*$
$n = 2$	$0.628743 < C^*$
$n = 3$	$0.632049 < C^*$
$n = 4$	$0.635817 < C^*$
$n = 5$	$0.636289 < C^*$
$n = 6$	$0.638918 < C^*$
$n = 7$	$0.639553 < C^*$
$n = 8$	$0.640063 < C^*$
$n = 9$	$0.641886 < C^*$
$n = 10$	$0.642005 < C^*$
$n = 11$	$0.643175 < C^*$
$n = 12$	$0.643566 < C^*$
$n = 13$	$0.643806 < C^*$
$n = 14$	$0.643859 < C^*$

$n = 15$	$0.644959 < C^*$
$n = 16$	$0.645582 < C^*$
$n = 17$	$0.645732 < C^*$

This author considers then the 'least squares method', amounting to assuming

$$u^{(n)} = \sum_{k=0}^{n} \mathrm{d}_k^{(n)} \varphi_k,$$

where $\varphi_0, \varphi_1, \ldots, \varphi_k, \ldots$ is the sequence introduced before, considered now as a complete system in $\mathfrak{L}^2(\Sigma)$. Fixing a 'weight' p on Σ such that $0 < q_0 \leqslant p(x) \leqslant q_1$ (q_0 and q_1 are some constants), the constants $\mathrm{d}_k^{(n)}$ are chosen by imposing the condition

$$\mathrm{I}^{(n)} = \int_\Sigma (1 - u^{(n)})^2 p \, d\sigma = \text{minimum}.$$

Suppose that the additional hypothesis has been made on Σ of the existence of a continuous vector $\mu(x)$, always penetrating into E, i.e. such that $\mu(x)\, \nu(x) \geqslant c_0 > 0$ for $x \in \Sigma$. Assume then $p = (\mu\nu)^{-1}$. Set:

$$\mathrm{C}^{(n)} = -\frac{1}{4\pi} \int_\Sigma \frac{\partial u^{(n)}}{\partial \nu} d\sigma = \mathrm{d}_0^{(n)}$$

and let $\{\alpha_{hk}^{(n)}\}$ $(h, k = 1, \ldots, n)$ be the inverse of the matrix

$$\left\{ \int_\Sigma \varphi_h \varphi_k \, p \, d\sigma \right\} \qquad (h, k = 1, \ldots, n).$$

Extending a former result of Payne and Weinberger [174], Sneider proves that

$$\frac{\mathrm{I}^{(n)}}{8\pi} \mathrm{H} - \sqrt{\frac{(\mathrm{I}^{(n)}\mathrm{H})^2}{64\pi^2} + \mathrm{I}^{(n)} \left(\frac{\mathrm{HC}^{(n)}}{4\pi} - \alpha_{00}^{(n)} \right)} \leqslant \mathrm{C} - \mathrm{C}^{(n)} \leqslant \tag{2.9.8}$$

$$\leqslant \frac{\mathrm{I}^{(n)}}{8\pi} \mathrm{H} + \sqrt{\frac{(\mathrm{I}^{(n)}\mathrm{H})^2}{64\pi^2} + \mathrm{I}^{(n)} \left(\frac{\mathrm{HC}^{(n)}}{4\pi} - \alpha_{00}^{(n)} \right)}.$$

H is a constant depending only on μ and on the geometry of Σ. It is explicitly computable. If, for instance, Σ is 'starlike' with respect to the origin, then we may assume $\mu(x) = x$ and H turns out to be equal to 1.

Applying the method to the case in which C is relative to the cube of side 2 and, hence, computing the corresponding $\mathrm{C}^{(n)}$, this author obtains, through (2.9.8), the following bounds on C^*:

n	Lower bound on C*	$\frac{C^{(n)}}{2}$	Upper bound on C*
0	0.570945	0.620286	0.684696
1	0.602571	0.635627	0.675290
2	0.611260	0.639609	0.672831
3	0.616169	0.641816	0.671469
4	0.621888	0.644344	0.669895
5	0.622549	0.644645	0.669745
6	0.626601	0.646428	0.668695
7	0.627522	0.646819	0.668434
8	0.628252	0.647154	0.668286
9	0.631077	0.648366	0.667538
10	0.631236	0.648434	0.667498
11	0.633083	0.649228	0.667028
12	0.633659	0.649486	0.666909
13	0.633994	0.649624	0.666813
14	0.634067	0.649657	0.666799
15	0.635785	0.650385	0.666357
16	0.636781	0.650809	0.666112
17	0.637002	0.650901	0.666002

Let us remark that the least square method has been applied by Gross [173], who, however, was unable to give rigorous estimates of the error.

Denote by $\Gamma(w_1, \ldots, w_m)$ the Green determinant, relative to the scalar product

$$\int_\Sigma vw(x \cdot \nu)^{-1}\, d\sigma$$

of m functions $w_1, \ldots, w_m$, defined in E (assumed starlike with respect to the origin).

It is shown in [176] that

$$4\pi \frac{\Gamma(\varphi_1, \ldots, \varphi_n)}{\Gamma(\varphi_0, \varphi_1, \ldots, \varphi_n)} \leqslant C. \tag{2.9.9}$$

The limit of the left-hand side as $n \to \infty$, is equal to C, the convergence being monotonic increasing.

Applying (2.9.9), which appeared not to be very effective, Sneider obtained the following lower bounds on C*:

$n = 0$	$0.428738 < C^*$
$n = 1$	$0.437189 < C^*$
$n = 2$	$0.446331 < C^*$
$n = 3$	$0.451946 < C^*$
$n = 4$	$0.458968 < C^*$
$n = 5$	$0.459949 < C^*$
$n = 6$	$0.465608 < C^*$
$n = 7$	$0.466846 < C^*$
$n = 8$	$0.468141 < C^*$
$n = 9$	$0.472366 < C^*$
$n = 10$	$0.472617 < C^*$
$n = 11$	$0.475624 < C^*$
$n = 12$	$0.476749 < C^*$
$n = 13$	$0.477261 < C^*$
$n = 14$	$0.477410 < C^*$
$n = 15$	$0.480351 < C^*$
$n = 16$	$0.482184 < C^*$
$n = 17$	$0.482567 < C^*$

Another approximation method for C is obtained in [176] considering the function $u^{(n)}$, as given by the least squares method, described above and assuming as an approximate value of C the quantity:

$$\Gamma^n = -\frac{1}{4\pi}\int_\Sigma u^{(n)}\frac{\partial u^{(n)}}{\partial \nu}\,d\sigma.$$

Set

$$K^{(n)} = \frac{1}{4\pi}\sum_{i,j}^{0,n} d_i^{(n)} d_j^{(n)} \int_\Sigma \frac{\partial \varphi_i}{\partial \nu}\frac{\partial \varphi_j}{\partial \nu}\frac{1}{p}\,d\sigma$$

$$Q^{(n)} = \frac{1}{4\pi}\sum_{h,k}^{0,n} b_h^{(n)} b_k^{(n)} \int_\Sigma \varphi_h \frac{\partial \varphi_k}{\partial \nu}\,d\sigma$$

where the $b_k^{(n)}$'s are obtained by solving the system:

$$\sum_{k=0}^{n} b_k^{(n)} \int_\Sigma \varphi_h\varphi_k\, p\,d\sigma = \int_\Sigma u^{(n)}\frac{\partial \varphi_h}{\partial \nu}\,d\sigma \qquad (h = 0, 1, \ldots).$$

The following inequalities hold:

$$\frac{I^{(n)}H}{4\pi} - \sqrt{\frac{(I^{(n)}H)^2}{16\pi^2} + \frac{I^{(n)}}{2\pi}(H\Gamma^{(n)} + K^{(n)} - Q^{(n)} - 4\pi\alpha_{00}^{(n)})} \leqslant \tag{2.9.10}$$

$$\leqslant C - \Gamma^{(n)} \leqslant \frac{I^{(n)}H}{4\pi} + \sqrt{\frac{(I^{(n)}H)^2}{16\pi^2} + \frac{I^{(n)}}{2\pi}(H\Gamma^{(n)} + K^{(n)} - Q^{(n)} - 4\pi\alpha_{00}^{(n)})}$$

Formula (2.9.10), applied to the cube of side 2, yields the following bounds:

n	Lower bound on C*	$\frac{\Gamma^{(n)}}{2}$	Upper bound on C*
0	0.565568	0.639611	0.743788
1	0.598951	0.650132	0.714527
2	0.608190	0.651997	0.705550
3	0.613240	0.653426	0.701623
4	0.619298	0.654606	0.696105
5	0.619916	0.654820	0.695728
6	0.624285	0.655559	0.691717
7	0.625171	0.655791	0.691047
8	0.625887	0.655973	0.690522
9	0.628911	0.656461	0.687779
10	0.629302	0.656512	0.687452
11	0.633181	0.656798	0.683728

An elegant method, yielding upper bounds on C, unfortunately proved to be ineffective by numerical experiments, is the following one. Denote by $\lambda^{(n)}$ the largest root of the secular equation

$$\det\left\{\int_\Sigma \varphi_h\, p\, d\sigma \int_\Sigma \varphi_k\, p\, d\sigma - \lambda \int_\Sigma \varphi_h \varphi_k\, p\, d\sigma\right\} = 0$$

$$h, k = 1, \ldots, n.$$

One has of course $\lambda^{(n)} \leqslant \lambda^{(n+1)}$. It turns out that

$$C \leqslant \frac{H}{4\pi}\left\{\int_\Sigma p\, d\sigma - \lambda^{(n)}\right\} \tag{2.9.11}$$

the right-hand side being convergent by decreasing to C, if Σ is starlike with respect to the origin and $\mu(x) \equiv x$.

The numerical results relative to C*, obtained through (2.9.11), are contained in the following table:

$n = 1$	$C^* < 0.930240356089430$
$n = 2$	$C^* < 0.929867082095430$
$n = 3$	$C^* < 0.929866124600518$
$n = 4$	$C^* < 0.929859070453049$
$n = 5$	$C^* < 0.929853877632029$
$n = 6$	$C^* < 0.929853875965554$
$n = 7$	$C^* < 0.929853542404137$
$n = 8$	$C^* < 0.929853191053968$
$n = 9$	$C^* < 0.929853191053324$
$n = 10$	$C^* < 0.929853191032676$
$n = 11$	$C^* < 0.929853177988055$
$n = 12$	$C^* < 0.929853163748639$
$n = 13$	$C^* < 0.929853163748631$
$n = 14$	$C^* < 0.929853163748342$
$n = 15$	$C^* < 0.929853163748340$
$n = 16$	$C^* < 0.929852712499163$

Sneider obtained two further error estimates for the least squares method; they read

$$\frac{H}{8\pi}\left\{I^{(n)} - \sqrt{I^{(n)}\int_\Sigma p\,d\sigma}\right\} \leqslant C - C^{(n)} \leqslant \frac{H}{8\pi}\left\{I^{(n)} + \sqrt{I^{(n)}\int_\Sigma p\,d\sigma}\right\} \tag{2.9.12}$$

$$\frac{HI^{(n)}}{8\pi} - \sqrt{\left[C^{(n)} + \frac{HI^{(n)}}{8\pi}\right]^2 - \frac{C^{(n)^2}\int_\Sigma p\,d\sigma}{\int_\Sigma p\,d\sigma - I^{(n)}}} \leqslant C - C^{(n)} \leqslant \tag{2.9.13}$$

$$\leqslant \frac{HI^{(n)}}{8\pi} + \sqrt{\left[C^{(n)} + \frac{HI^{(n)}}{8\pi}\right]^2 - \frac{C^{(n)^2}\int_\Sigma p\,d\sigma}{\int_\Sigma p\,d\sigma - I^{(n)}}}\,.$$

(2.9.12) yields the following bounds on C^*:

n	Lower bounds on C*	Upper bounds on C*
0	0.567845	0.687793
1	0.599215	0.678647
2	0.607938	0.676153
3	0.612896	0.674742
4	0.618709	0.673075
5	0.619372	0.672922
6	0.623505	0.671791
7	0.624456	0.671500
8	0.625189	0.671349
9	0.628103	0.670512
10	0.628267	0.670466
11	0.630176	0.669935
12	0.630762	0.669805
13	0.631114	0.669694
14	0.631187	0.669678
15	0.632976	0.669165
16	0.634013	0.668879
17	0.634245	0.668809

(2.9.13) yields the following bounds on C^*:

n	Lower bounds on C*	Upper bounds on C*
0	0.570945	0.684697
1	0.601572	0.676290
2	0.610040	0.674052
3	0.614842	0.672796
4	0.620462	0.671322
5	0.621104	0.671189
6	0.625095	0.670202
7	0.626011	0.669945
8	0.626720	0.669819
9	0.629527	0.669088
10	0.629685	0.669048
11	0.631522	0.668589
12	0.632088	0.668480

13	0.632426	0.668382
14	0.632497	0.668369
15	0.634216	0.667925
16	0.635213	0.667680
17	0.635436	0.667618

Finally, Sneider applied (2.9.4) to the computation of C^*. Denoting by x, y, z the coordinates of the point P of X^3 and by Σ the cube defined by the conditions: $0 \leqslant x \leqslant 1, 0 \leqslant y \leqslant 1, 0 \leqslant z \leqslant 1$, let Σ_δ be the cube concentric and homothetic to Σ, whose side length is $1 - 2\delta$ $\left(0 < \delta < \frac{1}{2}\right)$. Let $L_i^{(\delta)} (i = 1, \ldots, 12)$ be the edges of Σ_δ and let $V_j^{(\delta)} (j = 1, \ldots, \nu)$ be its vertices. Denote by ξ_δ the potential generated by a charge distribution of constant unit density on Σ_δ. Let η_δ be the potential relative to an analogous charge distribution on the edges of Σ_δ.

Let θ_δ be the potential generated by eight unit charges lying on the edges of Σ_δ and let, finally, χ be the potential due to a unit charge placed in the centre 0 of Σ. One has:

$$\xi_\delta = \int_{\Sigma_\delta} \frac{d_Q \sigma}{PQ} = \Phi(x, y, \delta, \delta) + \Phi(x, y, \delta, 1 - \delta) + \Phi(y, \delta, x, \delta)$$

$$+ \Phi(y, \delta, x, 1 - \delta) + \Phi(\delta, x, y, \delta) + \Phi(\delta, x, y, 1 - \delta),$$

where

$$\begin{aligned}
\Phi(\alpha, \beta, \gamma, w) &= (1 - \alpha) \log [\beta + \sqrt{(\alpha - 1)^2 + \beta^2 + (\gamma - w)^2}] \\
&+ \beta \log [1 - \alpha + \sqrt{(\alpha - 1)^2 + \beta^2 + (\gamma - w)^2}] \\
&- \frac{\beta}{|\beta|} |\gamma - w| \operatorname{arctg} \left[\frac{|\beta|}{|\gamma - w|} \frac{1 - \alpha}{\sqrt{(\alpha - 1)^2 + \beta^2 + (\gamma - w)^2}} \right] \\
&+ \alpha \log [\beta + \sqrt{\alpha^2 + \beta^2 + (\gamma - w)^2}] - \beta \log [-\alpha + \sqrt{\alpha^2 + \beta^2 + (\gamma - w)^2}] \\
&- \frac{\beta}{|\beta|} |\gamma - w| \operatorname{arctg} \left[\frac{|\beta|}{|\gamma - w|} \frac{\alpha}{\sqrt{\alpha^2 + \beta^2 + (\gamma - w)^2}} \right] \\
&- (1 - \alpha) \log [\beta - 1 + \sqrt{(\alpha - 1)^2 + (\beta - 1)^2 + (\gamma - w)^2}] \\
&- (\beta - 1) \log [1 - \alpha + \sqrt{(\alpha - 1)^2 + (\beta - 1)^2 + (\gamma - w)^2}] \\
&+ \frac{\beta - 1}{|\beta - 1|} |\gamma - w| \operatorname{arctg} \left[\frac{|\beta - 1|}{|\gamma - w|} \frac{1 - \alpha}{\sqrt{(\alpha - 1)^2 + (\beta - 1)^2 + (\gamma - w)^2}} \right] \\
&- \alpha \log [\beta - 1 + \sqrt{\alpha^2 + (\beta - 1)^2 + (\gamma - w)^2}]
\end{aligned}$$

$$+ (\beta - 1) \log [-\alpha + \sqrt{\alpha^2 + (\beta - 1)^2 + (\gamma - w)^2}]$$

$$+ \frac{\beta - 1}{|\beta - 1|} |\gamma - w| \operatorname{arctg}\left[\frac{|\beta - 1|}{|\gamma - w|} \frac{2}{\sqrt{\alpha^2 + (\beta - 1)^2 + (\gamma - w)^2}}\right].$$

$$\eta_\delta = \sum_{i=1}^{12} \int_{L_i^{(\delta)}} \frac{ds}{PQ} =$$

$$= \log\left[\frac{1 - \delta - x + \sqrt{(x - 1 + \delta)^2 + (y - \delta)^2 + (z - \delta)^2}}{\delta - x + \sqrt{(x - \delta)^2 + (y - \delta)^2 + (z - \delta)^2}}\right]$$

$$+ \log\left[\frac{1 - \delta - z + \sqrt{(x - 1 + \delta)^2 + (y - \delta)^2 + (z - 1 + \delta)^2}}{\delta - z + \sqrt{(x - 1 + \delta)^2 + (y - \delta)^2 + (z - \delta)^2}}\right]$$

$$+ \log\left[\frac{1 - \delta - x + \sqrt{(x - 1 + \delta)^2 + (y - \delta)^2 + (z - 1 + \delta)^2}}{\delta - x + \sqrt{(x - \delta)^2 + (y - \delta)^2 + (z - 1 + \delta)^2}}\right]$$

$$+ \log\left[\frac{1 - \delta - z + \sqrt{(x - \delta)^2 + (y - \delta)^2 + (z - 1 + \delta)^2}}{\delta - z + \sqrt{(x - \delta)^2 + (y - \delta)^2 + (z - \delta)^2}}\right]$$

$$+ \log\left[\frac{1 - \delta - x + \sqrt{(x - 1 + \delta)^2 + (y - 1 + \delta)^2 + (z - \delta)^2}}{\delta - x + \sqrt{(x - \delta)^2 + (y - 1 + \delta)^2 + (z - \delta)^2}}\right]$$

$$+ \log\left[\frac{1 - \delta - z + \sqrt{(x - 1 + \delta)^2 + (y - 1 + \delta)^2 + (z - 1 + \delta)^2}}{\delta - z + \sqrt{(x - 1 + \delta)^2 + (y - 1 + \delta)^2 + (z - \delta)^2}}\right]$$

$$+ \log\left[\frac{1 - \delta - x + \sqrt{(x - 1 + \delta)^2 + (y - 1 + \delta)^2 + (z - 1 + \delta)^2}}{\delta - x + \sqrt{(x - \delta)^2 + (y - 1 + \delta)^2 + (z - 1 + \delta)^2}}\right]$$

$$+ \log\left[\frac{1 - \delta - z + \sqrt{(x - \delta)^2 + (y - 1 + \delta)^2 + (z - 1 + \delta)^2}}{\delta - z + \sqrt{(x - \delta)^2 + (y - 1 + \delta)^2 + (z - \delta)^2}}\right]$$

$$+ \log\left[\frac{1 - \delta - y + \sqrt{(x - \delta)^2 + (y - 1 + \delta)^2 + (z - \delta)^2}}{\delta - y + \sqrt{(x - \delta)^2 + (y - \delta)^2 + (z - \delta)^2}}\right]$$

$$+ \log\left[\frac{1 - \delta - y + \sqrt{(x - \delta)^2 + (y - 1 + \delta)^2 + (z - 1 + \delta)^2}}{\delta - y + \sqrt{(x - \delta)^2 + (y - \delta)^2 + (z - 1 + \delta)^2}}\right]$$

$$+ \log\left[\frac{1 - \delta - y + \sqrt{(x - 1 + \delta)^2 + (y - 1 + \delta)^2 + (z - \delta)^2}}{\delta - y + \sqrt{(x - 1 + \delta)^2 + (y - \delta)^2 + (z - \delta)^2}}\right]$$

$$+\log\left[\frac{1-\delta-y+\sqrt{(x-1+\delta)^2+(y-1+\delta)^2+(z-1+\delta)^2}}{\delta-y+\sqrt{(x-1+\delta)^2+(y-\delta)^2+(z-1+\delta)^2}}\right]$$

$$\theta_\delta=\sum_{j=1}^{8}\frac{1}{\overline{PV_j^{(\delta)}}}=$$

$$=\frac{1}{\sqrt{(x-\delta)^2+(y-\delta)^2+(z-\delta)^2}}+\frac{1}{\sqrt{(x-1+\delta)^2+(y-\delta)^2+(z-\delta)^2}}$$

$$+\frac{1}{\sqrt{(x-1+\delta)^2+(y-1+\delta)^2+(z-\delta)^2}}+\frac{1}{\sqrt{(x-\delta)^2+(y-1+\delta)^2+(z-\delta)^2}}$$

$$+\frac{1}{\sqrt{(x-\delta)^2+(y-\delta)^2+(z-1+\delta)^2}}+\frac{1}{\sqrt{(x-1+\delta)^2+(y-\delta)^2+(z-1+\delta)^2}}$$

$$+\frac{1}{\sqrt{(x-1+\delta)^2+(y-1+\delta)^2+(z-1+\delta)^2}}+\frac{1}{\sqrt{(x-\delta)^2+(y-1+\delta)^2+(z-1+\delta)^2}}$$

Then one has:

$$\chi=\frac{1}{\overline{PQ}}=\frac{1}{\sqrt{\left(x-\frac{1}{2}\right)^2+\left(y-\frac{1}{2}\right)^2+\left(z-\frac{1}{2}\right)^2}}$$

Sneider has, therefore, considered a function of the type

$$v=c_1\xi_{\delta_1}+c_2\eta_{\delta_2}+c_3\theta_{\delta_3}+c_4\chi. \tag{2.9.14}$$

To apply (2.9.4), as far as the lower bound is concerned, this author chose $\delta_1=0$, $\delta_2=0.2$, $\delta_3=0.2$ and determined c_1, c_2, c_3, c_4 through a procedure analogous to the Daboni one, but divided any face of Σ in 10^6 equal squares (unlike Daboni, Sneider employed an electronic computer!). The results of the computation are:

$$c_1=0.22666579,\qquad c_2=-0.050491426,$$

$$c_3=0.0058826525,\quad c_4=-0.28692859.$$

Denoting by v_1 the function obtained by (2.9.4) in this way, the following tables show the numerical explorations performed to an increasingly small step, in order to give an upper estimate on v_1:

Table 11

Numerical values of the function v_1.

0.5						0.9954861
0.4					1.0005360	0.9982230
0.3				1.0037632	1.0029769	1.0015732
0.2			0.9957835	0.9976650	0.9974103	0.9967958
0.1		1.0094986	1.0019905	0.9991900	0.9992070	0.9992201
y = 0.0	0.9074603	0.9738776	0.9855558	0.9893105	0.9916116	0.9923863
x =	0.0	0.1	0.2	0.3	0.4	0.5

Table 12

Numerical values of the function v_1.

0.12					1.0070887
0.11				1.0086873	1.0079674
0.10			1.0094986	1.0091972	1.0086633
0.09		1.0092808	1.0095119	1.0094389	1.0091155
0.08	1.0078285	1.0086891	1.0091780	1.0093501	1.0092575
y = 0.07	1.0065342	1.0076729	1.0084302	1.0088599	1.0090135
x =	0.08	0.09	0.10	0.11	0.12

Table 13

Numerical values of the function v_1.

0.100						1.0094986
0.099					1.0095268	1.0095138
0.098				1.0095443	1.0095368	1.0095261
0.097			1.0095512	1.0095489	1.0095437	1.0095354
0.096		1.0095471	1.0095503	1.0095505	1.0095476	1.0095416
y = 0.095	1.0095318	1.0095407	1.0095464	1.0095489	1.0095484	1.0095448
x =	0.095	0.096	0.097	0.098	0.099	0.100

Table 14

Numerical values of the function v_1.

0.0971						1.00955107
0.0970					1.00955123	1.00955111
0.0969				1.00955129	1.00955130	1.00955127
0.0968			1.00955123	1.00955126	**1.00955130**	1.00955126
0.0967		1.00955115	1.00955123	1.00955126	1.00955127	1.00955129
y = 0.0966	1.00955089	1.00955103	1.00955112	1.00955119	1.00955127	1.00955127
x =	0.0966	0.0967	0.0968	0.0969	0.0970	0.0971

Table 15

Numerical values of the function v_1.

0.09687			1.00955136	1.00955136	1.00955136	1.00955133
0.09686		1.00955138	1.00955133	1.00955136	1.00955135	1.00955129
0.09685	1.00955130	1.00955132	1.00955136	1.00955135	1.00955135	1.00955132
0.09684	1.00955135	1.00955133	1.00955135	1.00955136	1.00955135	1.00955133
0.09683	1.00955133	**1.00955139**	1.00955130	1.00955133	1.00955136	1.00955133
0.09682	1.00955135	1.00955133	1.00955135	1.00955136	1.00955138	1.00955130
0.09681	1.00955132	1.00955132	1.00955133	1.00955133	1.00955136	1.00955132
0.09680	1.00955132	1.00955132	1.00955130	1.00955133	1.00955130	1.00955127
0.09679	1.00955132	1.00955132	1.00955130	1.00955130	1.00955132	1.00955132
0.09678	1.00955132	1.00955132	1.00955132	1.00955127	1.00955130	1.00955129
0.09677	1.00955129	1.00955132	1.00955133	1.00955133	1.00955127	1.00955130
0.09676	1.00955132	1.00955132	1.00955133	1.00955132	1.00955133	1.00955129
y = 0.09675	1.00955132	1.00955126	1.00955130	1.00955133	1.00955133	1.00955126
x =	0.09685	0.09686	0.09687	0.09688	0.09689	0.09690

Table 16

Numerical values of the function v_1.

0.096834	1.00955136	1.00955132	1.00955130	1.00955136	1.00955132	1.00955136
0.096833	1.00955136	1.00955136	1.00955130	1.00955133	1.00955136	1.00955135
0.096832	1.00955135	1.00955133	1.00955136	1.00955132	1.00955129	1.00955135
0.096831	1.00955135	1.00955133	1.00955135	1.00955133	1.00955130	1.00955133
0.096830	1.00955138	1.00955133	1.00955135	1.00955133	1.00955130	1.00955133
0.096829	1.00955135	1.00955138	1.00955133	1.00955136	1.00955135	1.00955133
0.096828	1.00955136	1.00955135	1.00955140	1.00955139	1.00955133	1.00955130
0.096827	1.00955130	1.00955133	1.00955136	1.00955135	1.00955133	1.00955136
0.096826	1.00955133	1.00955135	1.00955135	1.00955136	1.00955135	1.00955139
0.096825	1.00955138	1.00955135	1.00955133	1.00955135	1.00955136	1.00955135
0.096824	1.00955136	1.00955133	1.00955133	1.00955133	1.00955136	1.00955135
y = 0.096823	1.00955135	1.00955133	1.00955138	1.00955135	1.00955133	1.00955136
x =	0.096858	0.096859	0.096860	0.096861	0.096862	0.096863

The first inequality (2.9.4) gave, using v_1, the following lower bound on C^*:

$$0.653404 < C^*. \tag{2.9.15}$$

To obtain an upper bound on C^*, through (2.9.4), Sneider considered the function

$$v_2 = c_1 \xi_{\delta_1} + c_2 \eta_{\delta_2} + c_3 \theta_{\delta_3} + c_4 \chi + c_5 \theta_{\delta_4}$$

and assumed $\delta_1 = 0$, $\delta_2 = 0.01$, $\delta_3 = 0.001$, $\delta_4 = 0.1$, and

$$c_1 = 0.080995501, \qquad c_2 = 0.013296285, \qquad c_3 = 0.0013876246$$

$$c_4 = -0.025415031, \qquad c_5 = 0.0045256532.$$

The constants c_k have been computed through a procedure analogous to that employed for v_1.

The following tables show the numerical explorations, performed in order to give a lower estimate on v_2.

Table 17

Numerical value of the function v_2.

0.5						1.0006837
0.4					1.0005132	1.0005878
0.3				0.9998302	0.9999474	0.9999294
0.2			1.0012643	0.9989789	0.9981928	0.9980453
0.1		1.0101295	1.0023678	0.9972573	0.9963521	0.9963145
$y = 0.0$	1.0000279	0.9961552	0.9988287	0.9999280	1.0011649	1.0016915
$x =$	0.0	0.1	0.2	0.3	0.4	0.5

Table 18

Numerical value of the function v_2.

0.10				1.0101295
0.09			1.0095894	1.0098365
0.08		1.0088064	1.0091457	1.0093278
0.07	1.0082663	1.0084351	1.0086333	1.0087339
0.06	1.0083042	1.0082138	1.0082422	1.0082522
0.05	1.0087905	1.0084004	1.0082475	1.0081699
0.04	1.0100814	1.0093882	1.0090707	1.0089273
0.03	1.0127550	1.0118204	1.0113956	1.0112320
0.02	1.0173479	1.0163212	1.0158913	1.0157766
0.01	1.0175327	1.0166388	1.0163400	1.0163590
$y = 0.00$	0.9964644	0.9959447	0.9959174	0.9961552
$x =$	0.07	0.08	0.09	0.10

Table 19

Numerical value of the function v_2

0.010	1.0165020	1.0164665	1.0164356	1.0164089	1.0163863	1.0163674	1.0163520
0.009	1.0153961	1.0153628	1.0153340	1.0153094	1.0152888	1.0152719	1.0152585
0.008	1.0140125	1.0139815	1.0139549	1.0139325	1.0139142	1.0138993	1.0138880
0.007	1.0123706	1.0123421	1.0123179	1.0122979	1.0122817	1.0122692	1.0122601
0.006	1.0104996	1.0104737	1.0104521	1.0104345	1.0104208	1.0104106	1.0104039
0.005	1.0084330	1.0084098	1.0083909	1.0083759	1.0083647	1.0083571	1.0083527
0.004	1.0062034	1.0061831	1.0061669	1.0061547	1.0061462	1.0061411	1.0061393
0.003	1.0038380	1.0038208	1.0038076	1.0037982	1.0037925	1.0037901	1.0037910
0.002	1.0013538	1.0013397	1.0013296	1.0013232	1.0013204	1.0013208	1.0013244
0.001	0.9987467	0.9987359	0.9987289	0.9987257	0.9987258	0.9987292	0.9987356
y = 0.000	0.9958973	0.9958899	0.9958863	0.9958862	0.9958895	0.9958959	0.9959053
x =	0.083	0.084	0.085	0.086	0.087	0.088	0.089

Table 20

Numerical values of the function v_2

0.0010	0.9987276	0.9987272	0.9987269	0.9987265	0.9987263	0.9987260	0.9987258
0.0009	0.9984593	0.9984589	0.9984586	0.9984583	0.9984581	0.9984579	0.9984577
0.0008	0.9981890	0.9981887	0.9981884	0.9981881	0.9981879	0.9981877	0.9981876
0.0007	0.9979165	0.9979162	0.9979160	0.9979158	0.9979156	0.9979155	0.9979153
0.0006	0.9976415	0.9976414	0.9976412	0.9976410	0.9976408	0.9976407	0.9976407
0.0005	0.9973639	0.9973637	0.9973635	0.9973634	0.9973633	0.9973632	0.9973631
0.0004	0.9970829	0.9970827	0.9970826	0.9970824	0.9970824	0.9970823	0.9970823
0.0003	0.9967977	0.9967975	0.9967974	0.9967973	0.9967973	0.9967973	0.9967973
0.0002	0.9965069	0.9965068	0.9965067	0.9965066	0.9965066	0.9965066	0.9965067
0.0001	0.9962076	0.9962075	0.9962075	0.9962075	0.9962075	0.9962075	0.9962076
y = 0.0000	0.9958859	0.99588588	0.99588586	0.99588587	0.9958859	0.9958859	0.9958859
x =	0.0853	0.0854	0.0855	0.0856	0.0857	0.0858	0.0859

Using v_2, (2.9.4) yields the following upper bound on C^*:

$$C^* < 0.666394.$$

The last upper bound has been obtained by Sneider through (2.9.8) and the best lower bound through (2.9.4). It must be added that this result may be criticized because the result deduced from Table 15 and reported above is to be considered as an experimental result and not as one that has been mathematically achieved, no matter how accurately is the investigation performed in order to give an upper estimate on v_1. However, even if the lower bound given by (2.9.15) is not accepted, Sneider's results improve the former ones, since (2.9.7) yields in a completely rigorous way:

$$0.645732 < C^*.$$

Such an estimate was further improved by Tortorici [183] who, pushing the approximation by the method given by (2.9.7) up to $n = 29$, obtained, by very sharp computer calculations, the remarkable result:

$$0.6482812217 < C^*.$$

The writer ranks the memoir [176] by Sneider among the most advanced contributions made to an extremely difficult problem of numerical analysis. Besides the interesting results achieved, some experiments performed by this author represent a true example to the numerical analyst. For, people performing numerical computation, being sometimes unable to solve the hard analytical problems consisting in obtaining an error estimate, are content to verify that the procedures they employ yield approximations which keep 'stable' as the approximation goes further. Now, examine the results of the (lower) approximation method for C^* through (2.9.9) on page 177. The last result is obtained employing spherical harmonics up to those of degree 24 (*see* (2.9.5) and [176]) and keep in mind that the functions employed have all the cube symmetries).

Taking into account that there exist $2\nu + 1$ linearly independent spherical harmonics of degree ν, the introduction of as many spherical harmonics up to the degree m in the approximations is equivalent to having pushed the approximation up to the $(m + 1)^2$-th order.

In other words, if the functions employed in applying the method are taken to be orthonormalized, we have to sum a series of $(m + 1)^2$ terms. For $m = 24$ one has $(m + 1)^2 = 625$. Now the first significant figure 4 in the Table on page 177 kept stable from the first approximation up to the 625-th one, i.e. through as many as 140 approximations. However, it would be a mistake to think that the first three figures of C^* are 0.48!

Even more impressive is the example given by the upper estimate through (2.9.11) (*see* the Table on page 179). After 576 approximations one would end up with

$$C^* \simeq 0.92985. \tag{2.9.16}$$

Here are the approximations in which the various figures keep stable:

Figures	Approximation
9	First
2	36-th
9	36-th
8	36-th
5	100-th

The value given by (2.9.16) thus kept stable through 477 approximations! However, it is very far from the actual value of C^*.

10. Distribution of Electric Load near the Vertices and the Edges of a Cube

Let A be a bounded domain of the x, y, z space. Suppose that the boundary of A is a closed surface Σ. Let E be the set of all the points exterior to A.

It is well known, as was mentioned in the preceding Section, that the electrostatic potential of a load in equilibrium on Σ is obtained by solving the following exterior Dirichlet problem:

$$\frac{\partial^2 u}{\partial x^2} + \frac{\partial^2 u}{\partial y^2} + \frac{\partial^2 u}{\partial z^2} = 0 \text{ in E} \tag{2.10.1}$$

$$u = 1 \text{ on } \Sigma, \qquad u(\infty) = 0.$$

If one denotes by n the exterior normal to Σ (i.e. the normal pointing to the interior of E), the function

$$\delta(\mathrm{Q}) = -\frac{1}{4\pi}\left[\frac{\partial u}{\partial n}\right]_{\mathrm{Q}}$$

represents the density of the electric load on Σ.

Hence we have in every point P of E:

$$u(\mathrm{P}) = \int_{\Sigma} \delta(\mathrm{Q}) \frac{1}{|\mathrm{P}-\mathrm{Q}|}\, d\sigma_{\mathrm{Q}} \tag{2.10.2}$$

where $d\sigma$ is the measure of the surface element on Σ.

Although equation (2.10.2) is very classical in electrostatics, its mathematical justification can be considered achieved only in the case that Σ is a smooth surface, for instance in the case that Σ belongs to C^{1+h} $(0 < h \leqslant 1)$, i.e. Σ has a uniformly Hölder-continuous normal field (with Hölder exponent h).

For a surface which has singular points, like edges and vertices, one can only state that the electric load is represented by a countable additive set-function μ, defined on every Borel subset B of Σ. Then the potential u is represented by the Stieltjes integral

$$u(P) = \int_{\Sigma} \frac{d\mu_Q}{|P - Q|}.$$

We recalled in the preceding Section that recently it has been shown that, for a broad class of closed surfaces, the measure μ is absolutely continuous, i.e.

$$\mu(B) = \int_{B} \delta(Q)\, d\sigma_Q$$

and moreover $\delta \in \mathcal{L}^2(\Sigma)$ [176].

This class of closed surfaces can be roughly described by saying that A must be C^2-homeomorphic to the closure of a domain whose boundary is a closed polyhedral surface (*see* [176]). However, this result is not able to provide a precise indication of the behaviour of $\delta(Q)$ when Q tends to a point Q_0 lying on the set of the singular points of Σ.

It has been reported that this problem had been solved by Dirichlet in the case that Σ is a cube, since he succeeded in obtaining an analytic representation for δ.

Unfortunately, assuming that this is true, the result of Dirichlet was lost.(11)

The following theorem [195], [196] connects the asymptotic behaviour of δ and grad u, near the singular points of a cube, with an eigenvalue problem.

2.XIV. *Let* A *be the domain defined by the limitations*: $0 < x < 1$, $0 < y < 1$, $0 < z < 1$. *Let* Σ *be its boundary. Set* $\delta(x, y) \equiv \delta(Q)$, *for* $Q \equiv (x, y, 0)$, $\rho^2 = x^2 + y^2 + z^2$ *and let* R *be such that* $0 < R < 1$. *Consider the three segments*:

$$I_1 \equiv \{Q \equiv (x, y, z); 0 \leqslant x \leqslant 1, y = 0, z = 0\}$$

$$I_2 \equiv \{Q \equiv (x, y, z); 0 \leqslant y \leqslant 1, x = 0, z = 0\}$$

$$I_3 \equiv \{Q \equiv (x, y, z); 0 \leqslant z \leqslant 1, x = 0, y = 0\}.$$

For

$$Q \equiv (x, y, z) \in E \cup \Sigma - (I_1 \cup I_2 \cup I_3) \qquad 0 < \rho < R \tag{2.10.3}$$

we have

$$\delta(x, y) = \mathcal{O}\,[\rho^{\alpha - \frac{2}{3}} (\rho - x)^{-\frac{1}{6}} (\rho - y)^{-\frac{1}{6}}],$$

$$|\,[\text{grad}\, u]_Q| = \mathcal{O}\,[\rho^{\alpha - \frac{1}{2}} (\rho - x)^{-\frac{1}{6}} (\rho - y)^{-\frac{1}{6}} (\rho - z)^{-\frac{1}{6}}]$$

---

(11)On p.420 of [197] one can read: 'Bemerkung des Herrn Heinrich Weber: Kirchhoff hat in einem anfangs der siebenziger Jahre in Heidelberg geführten Gespräche dem Unterzeichneten erzählt, dass Dirichlet, wie er aus dessen eigenem Munde wisse, im Besitze der Lösung des Problems gewesen sei, die Verteilung der Statischen Electricität auf der Oberfläche eines leitenden Würfels zu bestimmen, und dass auch dies für uns, wie so vieles andere, verloren sei. Näheres darüber hat sich nicht ermitteln lassen. Strassburg, März 1897.' On the other hand, on p.363 of [198] a letter of Jacobi to Neumann is reported, where he writes: '... D.sagte mir neulich, er habe zu seiner Freude ein Problem gelöst, mit dem er sich lange beschäftigt hat, die Verbreitung der Elektricität auf einem rechtwinkligen Parallelepipedum, die Form der Entwicklungen schreitet nach einfachen, doppelten, dreifachen etc. Integralen fort...'.

where α is the positive constant defined as follows.

Let T *be the domain of the* ξ, η *cartesian plane, defined by the inequalities*:

$$\eta < 3^{\frac{1}{2}}\xi, \qquad \eta > 0 \qquad (\xi - 2^{\frac{1}{2}})^2 + \eta^2 < 3.$$

Let us consider on T *the following eigenvalue problem*:

$$\Delta_2 v + \frac{4\lambda v}{(\xi^2 + \eta^2 + 1)^2} = 0 \text{ in T} \tag{2.10.4}$$

$$v = 0 \text{ for } (\xi - 2^{\frac{1}{2}})^2 + \eta^2 = 3, \qquad v_\eta(\xi, 0) = 0, \tag{2.10.5}$$

$$3^{\frac{1}{2}} v_\xi(\xi, 3^{\frac{1}{2}}\xi) - v_\eta(\xi, 3^{\frac{1}{2}}\xi) = 0.$$

This problem has an increasing sequence of positive eigenvalues. If we denote by λ_1 *the smallest eigenvalue, we have*:

$$\alpha = \frac{(1 + 4\lambda_1)^{\frac{1}{2}} - 1}{2}.$$

The asymptotic formulae are the best which one can obtain. That means that one cannot replace $\mathcal{O}$ *by* o *when* P *tends to any point* $P_0 \in I_1 \cup I_2 \cup I_3$.

The last assertion of the theorem means that, no matter how P_0 has been fixed on $I_1 \cup I_2 \cup I_3$, the following statement is false:

Given arbitrarily $\epsilon > 0$, *there exists* $\sigma_\epsilon > 0$ *such that, for*

$$P \in E \cup \Sigma - (I_1 \cup I_2 \cup I_3), \qquad |P - P_0| < \sigma_\epsilon$$

we have:

$$|\delta(x, y)| \leqslant \epsilon \rho^{\alpha - \frac{2}{3}} (\rho - x)^{-\frac{1}{6}} (\rho - y)^{-\frac{1}{6}},$$

$$|\operatorname{grad} u| \leqslant \epsilon \rho^{\alpha - \frac{1}{2}} (\rho - x)^{-\frac{1}{6}} (\rho - y)^{-\frac{1}{6}} (\rho - z)^{-\frac{1}{6}}.$$

The proof of Theorem 2.XIV is rather delicate. The main idea consists of obtaining a series development of grad u valid for each Q satisfying condition (2.10.3). To this end, if we denote by $\tilde{S}$ the intersection of the unit sphere S of radius 1 centred on the origin (0, 0, 0) with $\bar{A}$ and we set $S^* = S - \tilde{S}$ (*see* Fig. 2.9), we shall consider the eigenvalue problem

$$Lv + \nu v = 0 \text{ on } S^*, \tag{2.10.6}$$

$$v = 0 \text{ on } \partial S^*$$

where L is the Laplace–Beltrami operator on S^*.

Let $\nu_1 < \nu_2 \leqslant .. \leqslant \nu_k \leqslant ...$ be the eigenvalues of the problem (2.10.6) and $\{v_k(\omega)\}$ a corresponding orthonormal system of eigenfunctions, ω is the point on S. The mentioned development is obtained by differentiation from the following:

$$u(\rho\omega) = 1 + \sum_{k=1}^{\infty} c_k \, \rho^{\alpha_k} v_k(\omega),$$

where the c_k's are suitable constants and

$$\alpha_k = \frac{(1 + 4\nu_k)^{\frac{1}{2}} - 1}{2}.$$

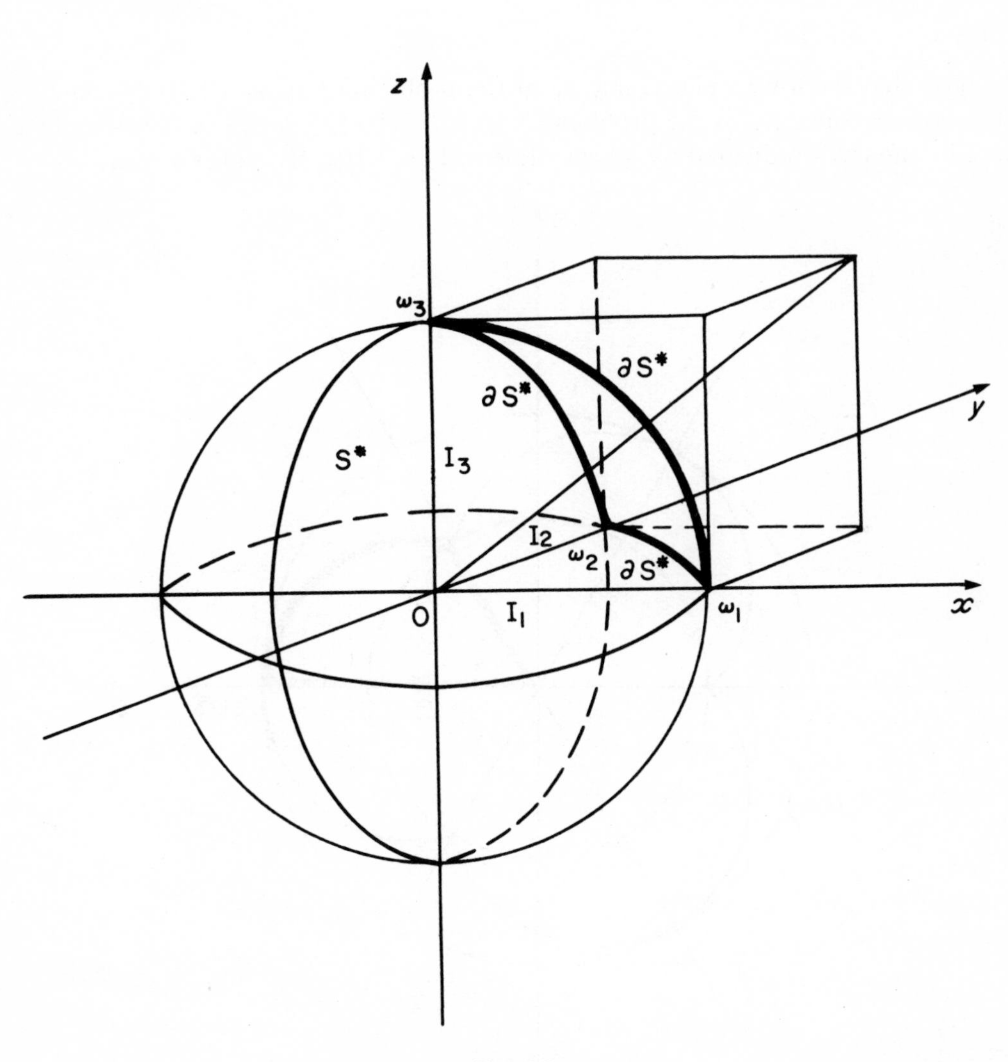

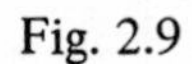
Fig. 2.9

The eigenvalue problem (2.10.6) is equivalent, by a suitable stereographic projection, to the one relative to the differential equation (2.10.4) in the domain D of the $\xi\eta$-plane defined by the inequalities

$$(\xi - 2^{\frac{1}{2}})^2 + \eta^2 < 3 \qquad \text{for} - 3^{\frac{1}{2}}\,\xi \leqslant \eta \leqslant 3^{\frac{1}{2}}\,\xi,$$

$$(\xi + 2^{-\frac{1}{2}})^2 + (\eta - \frac{3^{\frac{1}{2}}}{2})^2 < 3 \qquad \text{for} \quad \eta \geqslant 0, \eta \geqslant 3^{\frac{1}{2}}\,\xi,$$

$$(\xi + 2^{-\frac{1}{2}})^2 + (\eta + \frac{3^{\frac{1}{2}}}{2})^2 < 3 \qquad \text{for} \quad \eta \leqslant 0, \eta \leqslant -3^{\frac{1}{2}}\,\xi$$

(*see* Fig. 2.10) with the boundary condition

$$v = 0 \qquad \text{on } \partial D. \tag{2.10.7}$$

We have that the lowest eigenvalue ν_1 of the problem (2.10.4), (2.10.7) coincides with the lowest eigenvalue λ_1 of the problem (2.10.4), (2.10.5) relative to the domain T (whose boundary is indicated by a heavy line in Fig. 2.10). Moreover $\alpha = \alpha_1$.

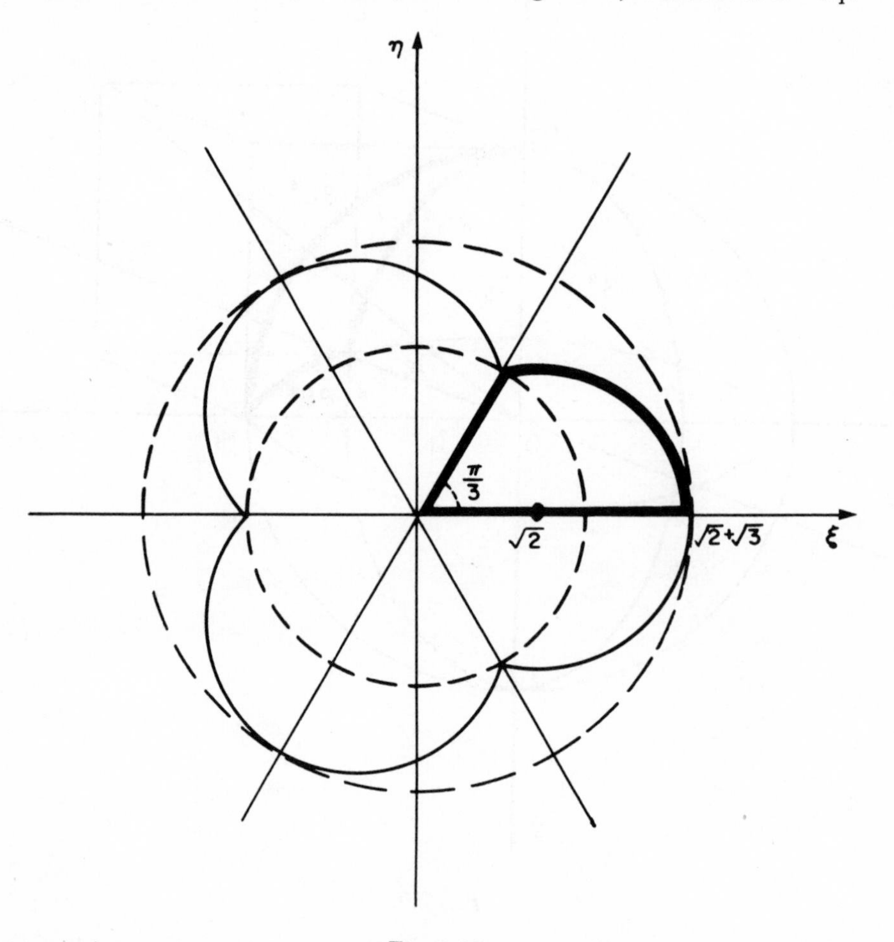

Fig. 2.10

Theorem 2.XIV reduces the determination of the asymptotic behaviour of δ and grad u, near the singular points of Σ, to the numerical solution of the eigenvalue problem (2.10.4), (2.10.5) as far as the first eigenvalue λ_1 is concerned. This eigenvalue problem is an *extremely difficult* one from the numerical point of view. The best rigorous bounds which have been obtained for λ_1 are the following:

$$0.62153 < \lambda_1 < 0.68025$$

and, in consequence,

$$0.4335 < \alpha < 0.4645. \tag{2.10.8}$$

The following theorems hold [195], [196]:

2.XV. *We have*

$$\delta(Q) \in \mathfrak{L}^p(\Sigma)$$

where p is any real number such that

$$1 < p < \min\left(\frac{2}{1-\alpha}, 3\right).$$

2.XVI. *We have*

$$|\operatorname{grad} u| \in \mathfrak{L}^p(E)$$

where p is any real number such that

$$\frac{3}{2} < p < \min\left(\frac{3}{1-\alpha}, 6\right).$$

In our opinion it is interesting that from rigorous results of numerical analysis, like the lower bound for α given by (2.10.8), we can get, through the Theorems 2.XV and 2.XVI, function theoretic results as the following:

$$\delta(Q) \in \mathfrak{L}^{3-\epsilon}(\Sigma)\ 0 < \epsilon \leqslant 2,$$

$$|\operatorname{grad} u| \in \mathfrak{L}^{6000/1133}(E).$$

Theorems 2.XIV, 2.XV and 2.XVI have been extended to more general surfaces [196], [199].

References

1 Nasta, M. Contributo al calcolo delle velocità critiche degli alberi motore. *Rend. Accad. Naz. Lincei, Cl. Sci. Fis. Matem. Nat.*, s. 6, **XII**, 209–216, 1930.

2 Gröbner, W. Über praktische Methoden zur Auflösung von Differentialgleichungen. *Jahres Bericht der Deutschen Math. Ver.*, **48**, 22–23, 1938.

3 Viola, T. Calcolo approssimato di autovalori. *Rend. Matem., Roma,* s. V, **II**, 71–106, 1941.

4 Collatz, L. Eigenwerte Probleme und ihre Numerische Behandlung. Chelsea Publ. Co., New York, 1948.

5 Kramers, H. A. Die Grundlagen der Quantentheorie–Quanten Theorie des Elektrons und der Strahlung. *Hand-und Jahrb. Chem. Phys.*, D Bd.I, *Theorien des Aufbauer der Materie* I, II, 200–201, Leipsiz: Akad. Ver., m.b.H., 1937.

6 Ghizzetti, A. Su un metodo di Picone per il calcolo degli autovalori e delle autofunzioni e su un'applicazione all'equazione di Schrödinger. *Proc. 8th Inter. Conf. Theor. Appl. Mech. Instanbul,* **II**, 1952.

7 Fichera, G. Results of recent experiments in the analysis of periods. Carried out in the Istituto Nazionale per le Applicazioni del Calcolo. *Nat. Bur. Stand. Appl. Math.* S. 29, 125–126, U. S. Gov. Print. Off., Washington, D.C., 1953.

8 Fichera, G. Su un metodo del Picone per il calcolo degli autovalori e delle autosoluzioni. *Ann. Matem. Pura Appl,* **40**, 239–259, 1955.

9 Bassotti, L. Costruzione di successioni approssimanti gli autovalori di una transformazione lineare e compatta. *Rend. Acc. Naz. Lincei, Cl. Sci. Fis. Matem. Nat.*, s. 8, **XXX**, 682–689, 1961.

10 Bassotti, L. Approssimazione globale degli autovalori di una transformazione lineare e compatta. *Rend. Acc. Naz. Lincei, Cl. Sci. Fis., Matem. Nat.*, s. 8, **XXX**, 828–832, 1961.

11 Osborn, J. E. Approximation of the eigenvalues of non-self-adjoint operators. *J. Math. Phys.*, **45**, 391–401, 1966.

12 Osborn, J. E. Approximation of the eigenvalues of a class of unbounded, non-self-adjoint operators. *SIAM J. Num. Anal.*, **4**, 1, 45–54, 1967.

13 Osborn, J.E. A method for approximating the eigenvalues of non-self-adjoint ordinary differential operators. *Mem. Acc. Sci. Torino, Cl. Sci. Fis., Matem. Nat.*, s. IV, **14**, 1–56, 1971.

14 Bramble, J. H. and Osborn, J.E. Rate of convergence estimates for non-self-adjoint eigenvalue approximation. *Matem. Res. Cent.,* Tech. Rep. 1232, 1972.

15 Schwartz, J. J. Perturbations of spectral operators and applications. I. Bounded perturbations. *Pacific J. Math.*, **4**, 415–458, 1954.

16 Gavurin, M. K. On estimates for eigenvalues and eigenvectors of a perturbated operator [in Russian]. *Dokl. Akad. Nauk SSSR,* s. 96, 1093–1095, 1954.

17 Rayleigh, Lord. The Theory of Sound. Macmillan, London & New York, **I**, 1894; **II**, 1896.

18 Ritz, W. Über eine neue Methode zur Lösung gewisser Variations-probleme der mathematischen Physik. *J. reine angew. Math.*, **135**, 1–61, 1908.

19 Ritz, W. Theorie der Transversalschwingungen einer quadratischen Platte mit freien Rändern. *Annal. Phys.*, Vierte Folge, **XXVIII**, 737–786, 1909.

20 Plancherel, M. Sur la méthode d'integration de Ritz. *Bull. Sci. Math.* s. II, **XLVII**, 376–384, 397–412, 1923; **XLVIII**, 12–48, 58–80, 93–109, 1924.

21 Fichera, G. Computation of eigenvalues and eigensolutions. *C.R. Congrès Int: Les Mathematiques de l'ingégneur*, Mons & Bruxelles, 58–69, Juin 1958.

22 Weinberger, H. F. Error bounds in the Rayleigh–Ritz approximation of eigenvectors. *Part. Diff. Eq. Cont. Mech.*, Ed. R. Langer, *Proc. Int. Conf.* (Math. Res. Cent., Univ. Wisconsin, 1960), The Univ. Wisconsin Press, Madison, 39–53, 1961.

23 Fischer, E. Über quadratische Formen mit reellen Koeffizienten. *Monats. Math. Phys.*, **16**, 234–249, 1905.

24 Weyl, H. Das asymptotische Verteilungsgesetz der Eigenwerte linearer partieller Differentialgleichungen. *Math. Annal*: **LXXI**, 441–479, 1911.

25 Weinstein, A. The intermediate problems and the maximum–minimum theory of eigenvalues. *J. Math. Mech.*, **12**, 235–245, 1963.

26 Fichera, G. Approximation and estimates for eigenvalues. *Proc. Symp. Num. Sol. Part. Diff. Eq.* (Univ. Maryland, May 1965). Ed. J. H. Bramble, Acad. Press, New York & London, 317–352, 1966.

27 Weinstein, A. Étude des spectres des équations aux dérivées partielles de la théorie des plaques élastiques . *Mémorial Sci. Math.* Gautier–Villars, Paris, 88, 65, 1937.

28 Aronszajn, N. Rayleigh–Ritz and Weinstein methods for approximation of eigenvalues. I. Operators in a Hilbert space. *Proc. Nat. Acad. Sci. USA*, **34**, 474–480, 1948.

29 Aronszajn, N. Approximation methods for eigenvalues of completely continuous symmetric operators. *Proc. Symp. Spectral Theory Diff. Probl.* (June–July 1950), Stillwater, Oklahoma, 179–202, 1951.

30 Zaremba, S. L'équation biharmonique et une classe remarquable des fonctions fondamentales harmoniques. *Bull. Int. Sci.* Cracovie, 147–196, 1907.

31 Aronszajn, N. and Weinstein, A. Sur la convergence d'un procédé variationnel d'approximation dans la théorie des plaques encastrées. *C. R. Sean. l'Acad. Sci.*, Paris, s. I, **204**, 96–98, 1937.

32 Fichera, G. Linear elliptic differential systems and eigenvalue problems. *Lecture Notes Math.*, 8, Springer, Berlin & New York, 1965.

33 Weinstein, A. and Stenger, W. Methods of intermediate problems for eigenvalues. *Math. Sci. Eng.*, **89**, Academic Press, New York & London, 1972.

34 Bazley, N. Lower bounds for eigenvalues. *J. Math. Mech.*, **10**, 289–307, 1961.

35 Weinberger, H. F. A theory of lower bounds for eigenvalues. *Tech. Note* BN–183, IFDAM, Univ. of Maryland, 1959.

36 Bazley, N. and Fox, D. W. Truncations in the method of intermediate problems for lower bounds to eigenvalues. *J. Res. Nat. Bur. Stand.*, **65**, 105–111, 1961.

37 Bazley, N. Lower bounds for eigenvalues with application to the helium atom. *Phys. Rev.*, (2), **120**, 144–149, 1960.

38 Bazley, N. and Fox, D.W. Lower bounds to eigenvalues using operator decomposition of the form B*B. *Arch. Rat. Mech. Anal.*, **10**, 352–360, 1962.

39 Bazley, N. and Fox, D.W. Methods for lower bounds to frequencies of continuous elastic systems. *Z. angew. Math. Phys.*, **XVII**, 1–37, 1966.

40 Kinoshita, T. Ground state of the helium atom. *Phys. Rev.*, (2), **105**, 1490–1502, 1957.

41 Gould, J. H. Variational methods for eigenvalue problems. 2nd edn., Univ. Toronto Press, 1966.

42 Weyl, H. Das asymptotische Verteilungsgesetz der Eigenschwingungen. *Rend. Circ. Matem. Palermo*, **39**, 1–50, 1915.

43 Courant, R. and Hilbert, D. *Methods of Mathematical Physics.* Interscience Publ., New York, vol. I, 1953.

44 Pleijel, A. A study of certain Green's functions with applications in the theory of vibrating membranes. *Ark. Math.*, **2**, 553–569, 1952.

45 Kac, M. Can one hear the shape of a drum? *Am. Math. Monthly,* **73**, 1–23, 1966.

46 McKean, H. P. and Singer, I. M. Curvature and the eigenvalues of the Laplacian. *J. diff. Geom.*, **I**, 43–69, 1967.

47 Courant, R. Über die Schwingungen eingespannter Platten. *Math. Z.*, **XV**, 195–200, 1922.

48 Pleijel, A. On the eigenvalues and eigenfunctions of elastic plates. *Comm. Pure App. Math.*, **III**, 1–10, 1950.

49 Carleman, T. Propriétes asymptotiques des fonctions fondamentales des membranes vibrantes. *8. Skandinav. Math. Kongress*, 34–44, 1934.

50 Carleman, T. Über die asymptotische Verteilung der Eigenwerte partieller Differentialgleichungen. *Ber. Verhandl. Sächs. Akad. Wissens. Leipzig, Mathema.-Nat., Wissens. Kl.*, **88**, 119–132, 1936.

51 Hardy, G. H. and Littlewood, J. E. Notes on the theory of series. XI. On Tauberian theorems. *Proc. Lond. Math. Soc.*, s. 2, **30**, 23–37, 1930.

52 Karamata, J. Neuer Beweis und Verallgemeinerung der Tauberschen Sätze, welche die Laplacesche und Stieltjessche Transformation betreffen. *J. reine angew. Math.*, **164**, 27–39, 1931.

53 Keldys, M. V. On the characteristic value and characteristic functions of certain classes of non-self-adjoint equations. *Dokl. Akad. Nauk SSSR (N.S.)*, **77**, 11–14, 1951.

54 Gårding, L. Dirichlet's problem for linear elliptic partial differential equations. *Math. Scand.*, **I**, 55–72, 1953.

55 Gårding, L. On the asymptotic distribution of the eigenvalues and eigenfunctions of elliptic differential operators. *Math. Scand.*, **I**, 237–255, 1953.

56 Browder, F. E. Le problème des vibrations pour un opérateur aux dérivées partielles self-adjoint et du type elliptique à coefficientes variables. *C. R. Sean. Acad. Sci.*, Paris, s. 2, **236**, 2140–2145, 1953.

57 Browder, F. E. Eigenfunction expansions with boundary conditions. *Am. J. Math.*, **LXXXVII**, 363–383, 1965.

58 Ehrling, G. On a type of eigenvalue problems for certain elliptic differential operators. *Math. Scand.*, **II**, 267–285, 1954.

59 Odhonoff, J. Operators generated by differential problems with eigenvalue parameter in equation and boundary condition (Thesis). *Comm. Sémin. Math. Univ. Lund,* **14**, 1959.

60 Bergendal, G. Convergence and summability of eigenfunction expansions connected with elliptic differential operators (Thesis). *Comm. Sémin. Math. Univ. Lund,* **15**, 5–63, 1959.

61 Agmon, S. On the eigenfunctions and on the eigenvalues of general elliptic boundary value problems. *Comm. Pure App. Math.,* **XV**, 119–147, 1962.

62 Agmon, S. Remarks on self-adjoint and semi-bounded elliptic boundary value problems. *Proc. Int. Symp. Lin. Spaces* (Jerusalem 1960), Jerusalem Acad. Press, Pergamon Press, 1–13, 1961.

63 Agmon, S. On the asymptotic distribution of eigenvalues of differential problems. *Ist. Naz. Alta Matem. Semin. 1962–1963,* **II**, Ediz. Cremonese, Roma, 556–573, 1965.

64 Agmon, S. Lectures on elliptic boundary value problems. Van Nostrand, New Jersey, New York, 1965.

65 Agmon, S. On kernels, eigenvalues, and eigenfunctions of operators related to elliptic problems. *Comm. Pure App. Math.,* **XVIII**, 627–663, 1965.

66 Agmon, S. and Kannai, Y. On the asymptotic behavior of spectral functions and resolvent kernels of elliptic operators. *Israel J. Math.,* **5**, 1–30, 1967.

67 Malliavin, P. Un théorème Taubérien avec reste pour la transformée de Stieltjes. *C. R. Sean. Acad. Sci.,* Paris, s. 3, **255**, 2351–2352, 1962.

68 Minakshisundaram, S. A generalization of Epstein zeta function. *Can. J. Math.,* **I**, 320–327, 1949.

69 Fichera, G. Sul calcolo degli autovalori. *Atti Conv. App. Analisi alla Fisica Matem.,* Cagliari–Sassari, 1964, Ediz. Cremonese, Roma, 37–55, 1965.

70 Fichera, G. Generalized biharmonic problem and related eigenvalue problems. *Blanch Annivers. Vol., Aerosp. Res. Lab.,* U.S. Air Force, Washington, D.C., 34–44, 1967.

71 Fichera, G. Sul miglioramento delle approssimazioni per difetto degli autovalori. *Rend. Acc. Naz. Lincei. Cl. Sci. Fis., Matem. Nat.,* s. 8, I and II, **XLII**, 138–145, 331–340, 1967.

72 Fichera, G. The problem of estimating eigenvalues when estimates in invariant subspaces are known. *Atti Acc. Sci. Torino,* **102**, 347–355, 1968.

73 Fichera, G. Invarianti ortogonali e autovalori degli operatori ellittici. *Rend. Semin. Matem. Fis. Milano,* **41**, 115–125, 1971.

74 Fichera, G. Further developments in the approximation theory of eigenvalues. *Symp. Numer. Sol. Part. Diff. Eq.* II. Ed. B. Hubbard, Acad. Press, New York & London, 243–252, 1971.

75 Fichera, G. Upper bounds for orthogonal invariants of some positive linear operators. *Rend. Ist. Matem. Univ. Trieste,* **I**, 1–8, 1969.

76 Fichera, G. The Neumann eigenvalue problem. *Applicable Analysis*, **3**, 213–240, 1973.

77 Fichera, G. Analytic problems concerning heterogeneous elastic media. *Proc. Symp. Cont. Mech. Related Problems of Analysis* (Tbilisi, 1971), **I**, 315–324, 1973.

78 Hellinger, E. Die Orthogonalinvarianten Quadratischen Formen von unendlich vielen Variablen. (Dissertation), Göttingen, 1907.

79 Hahn, H. Über die Integrale des Herrn Hellinger und die Orthogonal Invarianten der quadratischen Formen von unendlich vielen Veränderlichen. *Monats. Math. Phys.,* **23**, 161–224, 1912.

80 Plessner, A. I. and Rohlin, V. A. Spectral theory of linear operators [in Russian]. *Akad. Nauk SSSR: Moskov. Matem. Obščestvo.* Uspehi Matem. Nauk (Moscow & Leningrad), **II**, 71–191, 1946.

81 Halmos, P. R. Introduction to Hilbert space and the theory of spectral multiplicity. Chelsea Publ. Co., New York, **I**, 1951.

82 Trefftz, E. Über Fehlerschätzung bei Berechnung von Eigenwerten. *Math. Ann.*, **CVIII**, 595–604, 1933.

83 Weinstein, A. Some numerical results in intermediate problems for eigenvalues. *Proc. Symp. Numer. Sol. Part. Diff. Eq.*, Ed. J. H. Bramble, Acad. Press, New York & London, 167–191, 1966.

84 Weinstein, A. On the Sturm–Liouville theory and the eigenvalues of intermediate problems. *Numer. Math.*, **5**, 238–245, 1963.

85 Fichera, G. Approximation and estimates for eigenvalues. *Proc. Symp. Numer. Sol. Part. Diff. Eq.* (Univ. of Maryland, May 1965), Ed. J. H. Bramble, Acad. Press, New York & London, 317–352, 1966.

86 Fichera, G. Il calcolo degli autovalori. *Boll. Un. Matem. Ital.*, s. IV, **I**, 33–85, 1968.

87 Colautti, M. P. Sulle vibrazioni trasversali di una biella cuneiforme appoggiata agli estremi. *Rend. Acc. Naz. Lincei, Cl. Sci. Fis., Matem. Nat.*, s. 8, **XLIV**, 158–171, 1968.

88 Tricomi, F. Sulle vibrazioni trsversali di aste, specialmente di bielle di sezione variabile. *Ricerche di Ingegneria,* March–April 1936, **IV**, (2), 47–53, 1936.

89 Latzko, H. Der Wärmeübergang auf eine turbolente Flüssigkeit oder Gasstrom. *Z. angew. Math. Mech.*, **I**, 268–290, 1921.

90 Durfee, W. H. Heat-flow in a fluid with eddying flow (Readers Forum). *J. Aerosp. Sci.*, **23**, 1956.

91 Fettis, H. E. On the eigenvalues of Latzko's differential equation. *Z. angew. Math. Mech.*, **XXXVII**, 398–399, 1957.

92 Sansone, G. Sugli autovalori relativi ad un'equazione della conduzione del calore in un fluido soggetto a turbolenza. *Ann. Matem.*, **53**, 5–8, 1961.

93 Caligo, D. Calcolo del minimo autovalore relativo ad un'equazione della conduzione del calore in un fluido soggetto a turbolenza. *Rend. Acc. Naz. Lincei, Cl. Sci. Fis., Matem. Nat.*, s. 8, **XXXII**, 884–890, 1962.

94 Caligo, D. and Cotugno, N. Un calcolo di autovalori per una particolare equazione differenziale del secondo ordine. Quaderno 4, *Ist. Naz. Appl. Calcolo,* N. 640, Roma, 15–36, 1964.

95 Scarpini, F. Sul calcolo degli autovalori relativi ad una equazione differenziale inerente ad un problema di conduzione del calore. *Rend. Acc. Naz. Lincei, Cl. Sci. Fis. Matem. Nat.*, I & II, s. 8, **XXXVIII**, 53–60, 155–161, 1965.

96 Scarpini, F. Su un problema di autovalori relativo ad un'equazione differenziale singolare. *Centro Calcolo Ist. Matem. Univ. Roma,* 1–22, 1968.

97 Scarpini, F. Sulla distribuzione asintotica degli autovalori di un'equazione differenziale singolare. *Centro Calcolo Ist. Matem. Univ. Roma,* 1–10, 1968.

98 Scarpini, F. Sul calcolo degli autovalori connessi ad un problema di Genetica. *Rend. Matem. Roma,* s. V, **XXIV**, 80–107, 1965.

99 Colautti, M. P. Formule di maggiorazione per taluni invarianti ortogonali connessi con il calcolo rigoroso degli autovalori di un problema ai limiti. *Rend. Ist. Matem. Univ. Trieste,* **4**, 22–45, 1972.

100 Robert, D. Invariants orthogonaux pour certaines classes d'opérateurs. *J. Math. Pures Appl.*, t. 52, 81–114, 1973.

101 Grothendieck A. Théorie de Fredholm. *Bull. Soc. Math. France,* s. II, **LXXXIV**, 319–384, 1956.

102 Fichera, G. Existence theorems in elasticity. *Handbuch der Physik*, vol. VIa/2, Festkörpermechanik II, Springer Verl., Berlin, Heidelberg & New York, 347–389, 1972.

103 John, F. Plane waves and spherical means applied to partial differential equations. *Intersci. Tracts Pure Appl. Math.*, New York, no. 2, 1955.

104 Kelvin, Lord (Thomson, W.) Note on the integration of the equation of equilibrium of an elastic solid. *Cambridge and Dublin Math. J.*, **3**, 87–89, 1948.

105 Somigliana, C. Sulle equazioni della elasticità. *Ann. Matem. Pura Appl.*, 5. II, **17**, 37–64, 1889.

106 Fichera, G. Sull'esistenza e sul calcolo delle soluzioni dei problemi al contorno relativi allo equilibrio di un corpo elastico. *Ann. Sc. Norm. Sup. Pisa,* s. III, **IV**, 35–99, 1950

107 Bazley, N. W., Fox, D. W. and Stadter, J. T. Upper and lower bounds for the frequencies of rectangular clamped plates. *Z. angew. Mathem. Mech.*, **47**, 191–198, 1967.

108 De Vito, L., Fichera, G., Fusciardi, A. and Schaerf, M. Sul calcolo degli autovalori della piastra quadrata incastrata lungo il bordo. *Rend. Acc. Naz. Lincei, Cl. Sci. Fis. Matem. Nat.*, s. 8, **XL**, 725–733, 1966.

109 Clark, C. On the asymptotic formula for the eigenvalues of membranes and plates. *Boll. Un. Matem. Ital.*, s. IV, **III**, 191–196, 1970.

110 Schaerf, M. On the estimation of the eigenvalues of a circular plate clamped at the edge. *Mem. Acc. Naz. Lincei, Cl. Sci. Fis. Matem. Nat.*, s. 8, **IX**, 1–21, 1969.

111 Carrington, H. The frequencies of vibration of flat circular plates fixed at the circumference. *Lond. Edin. Dublin Philosoph. Mag. J. Sci.*, s. 6, **L**, 1261–1264, 1925.

112 Bassotti, L. Su un problema di autovalori per l'elasticità piana. *Riv. Matem. Univ. Parma,* s. 2, **VIII**, 259–289, 1967.

113 Bassotti, L. Calcolo numerico degli autovalori relativi al primo problema dell'elastostatica piana in un quadrato. *Riv. Matem. Univ. Parma,* s. 2, **IX**, 221–245, 1968.

114 Bassotti, L. Problemi quantitativi relativi al sistema alle derivate parziali della elastostatica piana. *Rend. Matem., Roma,* s. VI, **I**, 459–483, 1968.

115 Bassotti, L. Sottospazi invarianti per l'operatore della elasticità di un cubo. *Rend. Acc. Naz. Lincei, Cl. Sci. Fis. Matem. Nat.*, s. 8, **XLV**, 485–493, 1968.

116 Picone, M. Sur un problème nouveau pour l'équation linéaire aux dérivées partielles de la théorie mathématique classique de l'élasticité. *Second Colloq. Éq. Dér. Part.*, CBRM, Bruxelles, 9–11, 1954.

117 Lions, J. J. Contribution à un problème de Picone. *Ann. Matem. Pura App.*, s. IV, **XLI**, 201–219, 1956.

118 Campanato, S. Sul problema di Picone relativo all'equilibrio di un corpo elastico incastrato. *Ricerche Matem.*, **VI**, f. I, 125–149, 1957.

119 Colautti, M. P. Sui problemi variazionali di un corpo elastico incompressibile. *Mem. Acc. Naz. Lincei, Cl. Sci. Fis. Matem. Nat.*, s. 8, **VII**, 291–340, 1967.

120 Hodge, W. V. A Dirichlet problem for harmonic functionals, with applications to analytic varieties. *Proc. Lond. Math. Soc.*, s. 2, **36**, 257–303, 1934.

121 Colautti, M. P. Sulle vibrazioni assialmente simmetriche di una sfera elastica incompressibile. *Rend. Matem., Roma*, s. VI, **I**, 298–318, 1968.

122 Fichera, G. Invarianza rispetto al gruppo unitario e calcolo degli autovalori. *Symp. Math.*, Acad. Press, London & New York, **X**, 255–264, 1972.

123 Picone, M. Appunti di analisi superiore. *I Ediz.*, Rondinella, Napoli, 1940.

124 Colautti, M. P. Sul calcolo dei numeri di Betti di una varietà differenziabile, nota per mezzo di un suo atlante. *Rend. Matem., Roma,* s. V, **XXII**, 543–556, 1963.

125 Fichera, G. Spazi lineari di *k*-misure e di forme differenziali. *Proc. Int. Symp. Linear Spaces* (Jerusalem 1960), Jerusalem Acad. Press, Pergamon Press, 175–226, 1961.

126 De Vito, L., Sugli autovalori e sulle autosoluzioni di una classe di trasformazioni hermitiane. *Rend. Semin. Matem. Padova,* **XXV**, 144–176, 1956.

127 Wielandt, H. Die Einschliessung von Eigenwerten normalen Matrizen. *Math. Ann.*, **CXXI**, 595–604, 1949.

128 Bückner, H. Die praktische Behandlung von Integral Gleichungen. Springer Ver., Berlin, Göttingen & Heidelberg, 1952.

129 Fichera, G. Lezioni sulle trasformazioni lineari. *Ist. Matem. Univ. Trieste,* Ediz. Veschi, 1953.

130 Fox, L., Henrici, P. and Moler, C. Approximations and bounds for eigenvalues of elliptic operators. *SIAM J. Numer. Anal.*, **4**, 89–102, 1967.

131 Fichera, G. and Picone, M. Calcolo per difetto del più basso autovalore di un operatore ellittico del secondo ordine. *Rend. Acc. Naz. Lincei, Cl. Sci. Fis. Matem. Nat.*, s. 8, **XXX**, 411–418, 1961.

132 Sneider, M. A. Sul cerchio minimo contenente tutti gli autovalori di un operatore integrale di Fredholm con nucleo in L^2. *Rend. Acc. Naz. Lincei, Cl. Sci. Fis. Matem. Nat.*, s. 8, **XLVII**, 145–150, 1969.

133 Fichera, G. Sulla maggiorazione dell'errore di approssimazione nei procedimenti di integrazione numerica delle equazioni della Fisica Matematica. *Rend. Acc. Sci. Fis. Matem. Napoli*, s. IV, **XVII**, 138–145, 1950.

134 Fichera, G. Formule di maggiorazione connesse ad una classe di trasformazioni lineari. *Ann. Matem. Pura Appl.*, **36**, 273–296, 1954.

135 Caccioppoli, R. Limitazioni integrali per le soluzioni di un'equazione ellittica a derivate parziali. *Giorn. Matem. Battaglini*, s. IV, **LXXX**, f. **I**, 186–212, 1950.

136 Fichera, G. Methods of functional linear analysis in mathematical physics. *Proc. Int. Congr. Math.* (Amsterdam, 1954), **III**, sect. **V**, 216–228, 1956.

137 Davis, P. J. and Rabinowitz, P. Advances in orthonormalizing computation. *Advances in Computers*, Acad. Press, New York, vol. 2, 55–133, 1961.

138 Colautti, M. P. Sulla maggiorazione 'a priori' delle soluzioni delle equazioni e dei sistemi di equazioni differenziali lineari ordinarie del secondo ordine. *Le Matematiche*, **XI**, f. I, 8–99, 1956.

139 Beckenbach, E. F. and Bellman, R. Inequalities. Springer Ver., Berlin, Göttingen & Heidelberg, 1961.

140 Moretti, F. Alcune diseguaglianze relative alle funzioni armoniche nel complementare di un dominio limitato ed infinitesime all'infinito. *Boll. Un. Matem. Ital.*, s. III, **IX**, 190–198, 1954.

141 Caccioppoli, R. Sui teoremi d'esistenza di Riemann. *Ann. Sc. Norm. Sup. Pisa,* s. II, **VII**, 177–187, 1938.

142 Weyl, H. The method of orthogonal projection in potential theory. *Duke Math. J.*, **VII**, 411–444, 1940.

143 Lax, P. D. and Milgram, A. N. Parabolic equations. *Contributions to the Theory of Partial Differential Equations*, ed. L. Bers, S. Bochner and F. John, Princeton, New Jersey, 167–190, 1954.

144 Fichera, G. Alcuni recenti sviluppi della teoria dei problemi al contorno per le equazioni alle derivate parziali lineari. *Convegno Int. Eq. Lin. Der. Parz.*, Trieste, 25–28 August 1954, Ediz. Cremonese, Roma, 1955.

145 Faedo, S. Su un principio di esistenza nell'analisi lineare. *Ann. Sc. Norm. Sup. Pisa*, s. III, **XI**, 1–8, 1957.

146 Bramble, J. H. and Payne, L. E. Bounds for solutions of second-orders elliptic partial differential equations. *Contrib. to Diff. Eq.*, **I**, 95–127, 1963.

147 Bramble, J. H. and Payne, L. E. Some integral inequalities for uniformly elliptic operators. *Contrib. to Diff. Eq.*, **I**, 129–135, 1963.

148 Fichera, G. Premesse ad una teoria generale dei problemi al contorno per le equazioni differenziali. *Corsi INAM*, Ediz. Veschi, 1958.

149 Miranda, C. Formule di maggiorazione e teorema di esistenza per le funzioni biarmoniche di due variabili. *Giorn. Matem. Battaglini,* s. IV, **LXXVIII**, f. I, 97–118, 1948.

150 Fichera, G. Su un principio di dualità per talune formole di maggiorazione relative alle equazioni differenziali. *Rend. Acc. Naz. Lincei, Cl. Sci. Fis. Matem. Nat.*, s. 8, **XIX**, 411–418, 1955.

151 Fichera, G. Il teorema del massimo modulo per l'equazione dell'elastostatica tridimensionale. *Arch. Rat. Mech. Anal.* 7, 373–387, 1961.

152 Mikhlin, S. G. Multidimensional singular integrals and integral equations. Pergamon Press, 1965. [Russian original edn, 1962].

153 De Vito, L. Sopra una congettura di Mikhlin relativa ad una estensione delle diseguaglianze di M. Riesz alle superficie. *Rend. Matem., Roma*, s. V, **XXIII**, 273–297, 1964.

154 Fichera, G. Teoria assiomatica delle forme armoniche. *Rend. Matem., Roma*, s. V, **XX**, 147–171, 1961.

155 Fichera, G. Sulle equazioni differenziali lineari ellittico–paraboliche del secondo ordine. *Mem. Acc. Naz. Lincei, Cl. Sci. Fis. Matem. Nat.*, s. 8, sect. I, **V**, 1–30, 1956.

156 Fichera, G. On a unified theory of boundary value problems for elliptic–parabolic equations of second order. *Proceedings of a Symposium on Boundary Problems in Differential Equations*, ed. R. E. Langer (Madison, 1959), Univ. Wisconsin Press, 97–120, 1960.

157 Picone, M. Maggiorazione degli integrali delle equazioni totalmente paraboliche alle derivate parziali del secondo ordine. *Ann. Matem. Pura Appl.*, s. V, **VII**, 145–192, 1930.

158 Oleinik, O. and Radkewitch, E. Second order equations with non-negative characteristic form [in Russian]. *Itogi Nauka–Seriya Matematika*, Moscow 1971.

159 Lieberstein, H. M. A continuous method in numerical analysis applied to examples from a new class of boundary value problems. *Rend. Matem., Roma*, s. V, **XIX**, 347–378, 1960.

160 Lieberstein, M. H. A course in numerical analysis. Harper & Row, New York, Evaston & London, 1968.

161 Fichera, G. Sul concetto di problema 'ben posto' per un'equazione differenziale. *Rend. Matem., Roma*, s. V, **XIX**, 95–121, 1960.

162 Kötter, F. Über die Torsion des Winkeleisens. *Sitzungsber. König. Preuss. Akad. Wissens.*, **XXXIII**, 1908.

163 Trefftz, E. Über die Wirkung einer Abrundung auf die Torsionsspannungen in der inneren Ecke eines Winkeleisens. *Z. angew. Math. Mech.*, **2**, 263–267, 1922.

164 Tricomi, F. Nuovi risultati nel problema della trave sollecitata di taglio. *Ricerche di Ingegneria*, maggio-giugno, A. II, nr. 3, 93–104, 1934.

165 Picone, M. Sulla torsione di un prisma elastico cavo secondo la teoria di Saint-Venant. *Ann. Soc. Polon. Math.*, **XX**, 347–372, 1947.

166 Fichera, G. Sulla torsione elastica dei prismi cavi. *Rend. Matem., Roma*, s. V, **XII**, 163–176, 1953.

167 De Schwartz, M. J. Über das Verhalten der Torsionsfunktion in der Nähe von Einspringenden Ecken massiver und hohler Stäbe. *Österr. Ingenieur-Archiv, Wien,* Springer Ver. 8, **7**, 88–100, 1935.

168 Greenspan, D. Resolution of classical capacity problems by means of a digital computer. *Can. J. Phys., Ottawa,* **44**, 2605–2614, 1966.

169 Pólya, G. Estimating electrostatic capacity. *Amer. Math. Monthly,* **54**, 201–206, 1947.

170 Pólya, G. and Szegö, G. Isoperimetric inequalities in mathematical physics. *Ann. Math. Studies,* nr. 27, Princeton, 1951.

171 Pólya, G. and Szegö, G. Inequalities for the capacity of a condenser. *Amer. J. Math.*, **LXVII**, 1–32, 1945.

172 McMahon, J. Lower bounds for the electrostatic capacity of a cube. *Proc. Roy. Ir. Acad.*, sect. A, **55**, 133–167, 1953.

173 Gross, W. Sul calcolo della capacità elettrostatica di un conduttore. *Rend. Acc. Naz. Lincei, Cl. Sci. Fis. Matem. Nat.*, s. 8, **XII**, 496–506, 1952.

174 Payne, L. E. and Weinberger, H. F. New bounds in harmonic and biharmonic problems. *J. Math. Phys.*, **XXXIII**, nr. 4, 291–307, 1955.

175 Daboni, L. Applicazione al caso del cubo di un metodo per il calcolo per eccesso e per difetto della capacità elettrostatica di un conduttore. *Rend. Acc. Naz. Lincei, Cl. Sci. Fis. Matem. Naz.*, s. 8, **XIV**, 461–466, 1953.

176 Sneider, M. A. Sulla capacità elettrostatica di una superficie chiusa. *Mem. Acc. Naz. Lincei, Cl. Sci. Fis. Matem. Nat.*, s. 8, **X**, 97–215, 1970.

177 Diaz, J. B. Some recent results in linear partial differential equations. *Convegno Int. Eq. Lin. Der. Parz.* (Trieste, 25–28 August 1954), Ediz. Cremonese, Roma, 1955.

178 Diaz, J. B. Upper and lower bounds for quadratic integrals and, at a point, for solutions of linear boundary value problems. *Proc. Symp. Bound. Prob. Diff. Eq.*, ed. R. E. Langer, (Madison 1959), Univ. Wisconsin Press, 97–120, 1960.

179 Diaz. J. B. Upper and lower bounds for the torsional rigidity and the capacity, derived from the inequality of Schwarz. *Centro Int. Matem. Estivo* (*C.I.M.E.*), II ciclo, Chieti, 1–12, 1962.

180 Protter, M. and Weinberger, H. F. On the capacity of composite conductors. *J. Math. Phys.*, **XLIV**, 375–383, 1965.

181 Parr, W. E. Upper and lower bounds for the capacity of the regular solids. *J. Soc. Industr. App. Math.*, **9**, 334–386, 1961.

182 Sneider, M. A. Calcolo automatico di polinomi armonici omogenei aventi le simmetrie del cubo. *Calcolo*, **5**, f. IV, 557–580, 1968.

183 Tortorici, M. Analisi numerica di un algoritmo con notevoli esigenze di precisione. *Atti Acc. Sci. Lett. Arti Palermo*, Parte Prima: Scienze (4), **30**, 389–454, 1969/1970.

184 Cassisa, C. Maggiorazione di taluni invarianti ortogonali relativi ad un operatore differenziale ordinario del secondo ordine con condizioni ai limiti di tipo generale. *Rend. Ist. Matem. Univ. Trieste,* **VI**, 1–27, 1974.

185 Romano, M. On a class of buckling problems in the theory of elastic structures. *Rend. Matem., Roma*, s. VI, **9**, 215–234, 1976.

186 Bassotti, L. Sottospazi invarianti per operatori differenziali lineari a coefficienti costanti. *Rend. Circ. Matem. Palermo,* s. II, **XXII**, 157–184, 1973.

187 Bassotti, L. Sulle decomposizioni di alcuni spazi funzionali in somma diretta di sottospazi mutuamente ortogonali. *Rend. Acc. Naz. Lincei, Cl. Sci. Fis. Matem. Nat.*, s. VIII, **LVI**, 13–21, 1974.

188 Smith, G. F. Projection operators for symmetric regions. *Arch. Rat. Mech. Annal.*, **54**, 161–174, 1974.

189 Fichera, G. Approximation of the eigenvectors of a positive compact operator and estimate of the error. *Ann. Matem. Pura App.*, s. IV, **CVIII**, 1976.

190 Ostrowski, A. M. Integral inequalities. *Corso C.I.M.E. 1970 su 'Funct. Eq. Inequalities'*, Ediz. Cremonese, Roma, 1971.

191 Fichera, G. and Sneider, M. A. Un problema di autovalori proposto da Alexander M. Ostrowski. *Rend. Matem., Roma*, s. VI, **8**, 201–224, 1975.

192 Fichera, G. Osservazioni e risultati relativi al calcolo degli autovalori di taluni operatori positivi. *Boll. Un. Matem. Ital.*, s. 4, **11**, 430–443, 1975.

193 Oleinik O. A. and Radkevitch, E. V. Second order equations with non-negative characteristic form. *Amer. Math. Soc.*, Plenum Press, New York & London, 1973.

194 Oleinik, O. A. and Radkevitch, E. V. Equazioni del secondo ordine con forma caratteristica non negativa. Ediz. Veschi, Roma, 1975.

195 Fichera, G. and Sneider, M. A. Distribution de la charge électrique dans le voisinage des sommets et des arêtes d'un cube. *C. Rend. Acc. Sci. Paris*, s. A, **278**, 1303–1306, 1974.

196 Fichera, G. Comportamento asintotico del campo elettrico e della densità elettrica in prossimità dei punti singolari della superficie conduttore. *Rend. Sem. Matem. Univ. Politec. Torino,* **32**, 111–143 [Russian transl.: *Uspehi Matem. Nauk* **XXX**, 105–124].

197 Dirichlet, G. L. *Werke.* Herausgegeben auf Veranlassung der Königlich Preussischen Akademie der Wissenschaften von *L. Kronecker* forgestz von *L. Fuchs*, vol. 2, Georg Reimer, Berlin, 1897.

198 Koenigsberger, L. *Carl Gustav Jacob Jacobi.* Teubner, Leipzig, 1904.

199 Fichera, G. Asymptotic behaviour of the electric field near the singular points of the conductor surface. *Rend. Acc. Naz. Lincei, Cl. Sci. Fis., Matem. Nat.* s. VIII, **LX**, 13–20, 1976.

200 Castellani Rizzonelli, P. On the first boundary value problem for the classical theory of elasticity in a three-dimensional domain with a singular boundary. *J. Elasticity,* **3**, 225–259, 1973.

Author Index